Intimations of Immortality
Do You Want to Live Forever?

Gerald Leinwand

President Emeritus
Western Oregon University

AuthorHouse™
1663 Liberty Drive
Bloomington, IN 47403
www.authorhouse.com
Phone: 1-800-839-8640

© *2010 Gerald Leinwand. All rights reserved.*

No part of this book may be reproduced, stored in a retrieval system, or transmitted by any means without the written permission of the author.

First published by AuthorHouse   10/1/2010

ISBN: 978-1-4520-7328-6 (sc)
ISBN: 978-1-4520-7329-3 (hc)
ISBN: 978-1-4520-7330-9 (e)

Library of Congress Control Number: 2010912883

Printed in the United States of America

This book is printed on acid-free paper.

Because of the dynamic nature of the Internet, any Web addresses or links contained in this book may have changed since publication and may no longer be valid. The views expressed in this work are solely those of the author and do not necessarily reflect the views of the publisher, and the publisher hereby disclaims any responsibility for them.

To newlyweds Erica Opper and Per Midboe
May all your wishes come true, but may you be careful of what you wish for.

Musings on Immortality.

My grandfather died when he was sixty-five; my father died at seventy. At eighty-nine, I am the eldest of three brothers, all of us older than my father was at his death. Given a reasonable degree of vitality; however, I would like to live forever. Wouldn't you? I would like to be at my grandchildren's and great-grandchildren's weddings. Wouldn't you? I would like to see how it all comes out in the end of time for my family, my country, the world.

Wouldn't it be great if the intellectual giants—Socrates, Plato, Aristotle, Maimonides, Descartes, Rousseau, Voltaire, Copernicus, Galileo, Francis Bacon, Milton, Newton, Leonardo da Vinci, Michelangelo, Titian, Raphael, Hegel, Goethe, Gertrude Stein, Emily Dickinson, Harriet Beecher Stowe, Clara Barton, James Joyce, Robert Frost, Albert Einstein, to mention only a few—could be with us today? Many of them achieved their best work in literature, art, philosophy, science, or mathematics at advanced ages. Wouldn't the world be a better place if the wisdom of these talented people would still be around to set us straight?

During the European Middle Ages, the church looked askance at medical intervention as tampering with the will of God. The faithful, according to this view, must shun medical care and rely instead on prayer lest God's will be thwarted. Monks and regular clergy were forbidden by any number of church councils from administering or seeking medical care. Behind this view was that the body was unimportant as long as the soul was saved.

In our own time, when fairy tales of men and women who lived hundreds or even thousands of years still intrigue us, is the pursuit of eternal life pornographic as some suggest? Are not the greats of medicine who look for cures for catastrophic illness like cancer, heart disease, or AIDS, not also in the business of extending the life span of men and women? When August von Wassermann gave the world a diagnostic test for syphilis, when Jonas Salk developed the polio vaccine, were these men not engaged in a search for immortality?

What will be remembered most of the twenty-first century is that a conscious effort was made, not merely to treat disease, to prolong life, but to confront and perhaps conquer death? "Death be not proud," wrote John Donne. He was right.

While searching for immortality we are baffled by age. My first wife of forty-three years died of rampaging breast cancer, my second wife of six years, died of virulent brain cancer. Why was my first wife denied the golden years of age just as our children were out on their own and very successful in their career pursuits? Why was my second wife, a blondeblond bombshell, stricken just when she was at the top of her games of golf and tennis? Why was she denied the respite in Florida for which we had assiduously planned? How does one make sense as to why we are attacked by Parkinson's and Alzheimer's, by heart disease and cancer? Why the Holocaust? Why the terror of 9/11?

The paradox of immortality is that while we may live longer, we do not banish age and its ailments. Yet, the search may be its own reward by giving the aged, despite their ailments, a sense of direction, a purpose, a focus, and meaning often absent in later years. Thus, the young may see things as they are and ask "Why?" The old may dream of things that never were and ask "Why not?"

Gerald Leinwand

PROLOGUE

Happy Birthday

The first African American pitcher in the major leagues, the talented Satchel Paige, a member of baseball's National Hall of Fame, began his career on the mound at age seventeen and continued to pump his fastball at opposing batters with singular regularity over a period of several decades. His date of birth cannot be determined with accuracy, but 1905 is generally accepted. He pitched his first professional game in 1926, in the Negro Leagues, and his last when he was about sixty in 1965. When reporters with some wonderment asked him, "How old are you?" his quick reply was, "How old would you be if you didn't know how old you was?"[1]

In his new biography of Satchel Paige, journalist Larry Tye asserts that Satchel's age is the most argued statistic in sports. In 1934, the colored *Baseball and Sports* monthly reported that Satchel was born in 1907. According to the Associated Press, he was born in 1901, according to *Time*, he was born in 1903, and *the Post, the New York Times,* and the *Sporting News* report the year of his birth as 1908. Satchel's mother remembers the year of his birth as 1904."[2]

During much of human history, most people didn't know how old they were. The United States, "before the mid-nineteenth century was not a country where age played a vital part in people's everyday lives and associations."[3] Most people died of disease, famine, plague, accident, war, and by our standards, did not grow old. Since formally recording the exact date of a child's birth was not uniformly made, the birthdays of ordinary people could not be celebrated. The admonition to young and old alike "to act your age," was governed by convention, not the calendar.

Among toddlers, there are the terrible twos. At age five, children often begin their kindergarten and elementary schooling. At about eight years of age, Catholic children have their First Holy Communion. At age thirteen, a Jewish boy or girl may undergo the ceremony of bar mitzvah or bat mitzvah as they are formally accepted as grown members of the Jewish people. At about age fourteen, teenagers begin high school, a sweet sixteen party becomes an important social event for that age cohort, and a learning permit or a license to drive a car becomes an essential rite of passage. At age eighteen, young women and men can vote, join the military, and in some states, this age group may drink alcoholic beverages. Debutantes and wannabe debutantes come out. At about the same time, some of this group attends college and parents experience an empty nest syndrome.

By age twenty-two, young men and women are expected to begin work, or through further study, become prepared to do so. While less exact, marriage is expected to occur between the ages of twenty-five or thirty. "Forty is the old age of youth; fifty is the youth of old age."[4] The old retire at sixty, and Medicare and Social Security kick-in at about sixty-two, at which time many are also entitled to senior citizen discounts at movies or athletic events and on public transportation.

In 1893, sisters Mildred J. Hill and Patty Smith Hill from Louisville, Kentucky, published a song for kindergarten children in Chicago. A third sister, Jessica M. Hill claimed she also collaborated. The song, "Good Morning to All" morphed to become, "Happy Birthday to You," "America's most frequently sung musical piece."[5] By 1934, Irving Berlin and Moss Hart featured the song in the hit musical, *As Thousands Cheer*. But what had happened to make a simple tune so ubiquitous a hit?

As the precise date of birth of newborn infants became uniformly recorded, a birthday could be known with precision and celebrated with joy. It was not long before a birthday greeting card became commonplace. From Kansas City in 1910, Joyce Clyde Hall began the manufacture of birthday, Valentine, Easter, and Christmas cards. Before long, the Hallmark Card had a near monopoly of the field.

Birthday cards and popular melodies demonstrate an awareness of age, reveal a preference for youth, express regret for a youth's passing, and anticipate the bitter prospect of death. One such card declared on one side, "Don't feel old. We have a friend your age." On the reverse side, "and on good days, he can still feed himself." Or, on one side of another card, "You have just turned thirty." On the inside, "You will never have fun for the rest of your life." Or on one more, on the front side, "There are some things worse than birthdays …" On the reverse side, "like nuclear war."[6]

The young often pity, tolerate, or amuse the elderly even as they wait impatiently for their inheritance. While there is some nostalgic respect for age, few of us truly look forward to those big birthdays, the birthday cards that assure us that life begins at forty, or is it fifty? Or perhaps sixty?

Gerald Leinwand
Birthday: 8/27/21

Chapter 1

I Don't Remember Growing Older

The financier Bernard Baruch insisted, "Old age is always fifteen years older than I am." The popular French singer, Maurice Chevalier, responded to a query about how it felt to be old, "It isn't bad when you consider the alternative."[8] When the aged former president, Thomas Jefferson, was asked if he would choose to live life over again he said yes, but only between the ages of twenty-five and sixty. Thereafter, he wrote, "The powers of life are sensibly on the wane, sight becomes dim, hearing dull, memory constantly enlarging its frightful blank and parting with all we have every limb, and so faculty after faculty quits us, and where then is life?"[9]

In her book, *The Coming of Age,* Simone de Beauvoir reminds us, "There are no initiation rites"[10] that herald the coming of age. "Nothing should be more expected than old age; nothing is more unforeseen."[11] Except for the very young who look forward to growing older because of the privileges and opportunities for independence they believe age implies, most people, unconsciously think that while others may die, they are immortal.

In a study of 1700 people between the ages of twenty to eighty, some of the eighty year olds failed to perceive themselves as old.[12] For the young, we have schools, organized sports, camps, opportunity for music, skiing, or ice skating lessons. We prepare the young systematically for adult roles, but the elderly have to make their own preparations for growing old. As a song in the musical *Fiddler on the Roof* suggests, "I don't remember growing older!"

Becoming old is nearly always a shock.

Life Expectancy

Life expectancy at birth increased by thirty years in the last century, and according to most demographers, significant increases may be expected to continue.[13] Tables published by the National Center for Health Statistics, show that life expectancy at birth was 47.3 years in 1900, rose to 68.2 by 1950 and reached 77.3 in 2002. "Americans turning 65 in 2005 can expect to live on average until they are 83, four and a half years longer than the typical 65-year-old could expect in 1940. By 2040, the average 65-year-old will live to about 85. "[14] When President Franklin Roosevelt signed Social Security legislation into law in 1935, the life expectancy of a sixty year old man was twelve and a half years. Today, the life expectancy is sixteen years. Today, an average American man who retired at the age of fifty-five will spend 32 percent of his life enjoying his retirement years.

Some demographers, however, such as S. Jay Olshansky of the University of Illinois insist "that life expectancy has reached a ceiling, and that there is nothing on the horizon to indicate that life style changes, surgical procedures, vitamins, anti-oxidants, hormones, or techniques of genetic engineering have the capacity to repeat the gains in life expectancy that were achieved in the 20th century."[15] Yet, in the matter of aging, it is better to bet with the optimists and to prepare for the further "graying of America."

The Baby Boomers

Nine months after the end of World War II, live births in the United States surged from 222,721 in January of 1946 to 233,452, by May of that year. That October, an astonishing 339,499 babies were born the United States. By the end of the 1940s, 32 million babies had been born compared with 24 million in the 1930s. Between 1945 and 1964, 76 million babies had been born in the United States. Thus, "the great postwar baby boom had begun in a torrent of diapers, dishes, and debts."[16] Those infants born between 1946 and 1965 have been described as the "baby boomer" generation.

While it availed them little, the older generation deplored the spending habits and the sex habits of the boomers. The baby boomers in America moved to the suburbs where the automobile and television dominated. Women were expected to marry, but many began to work outside the home in careers of their own. Many boomers protested the Vietnam War, but men and many women went off to fight. The rate of divorce grew, as divorce was no longer looked upon as social stigma. By 2006, the baby boomers turned sixty and some became grandparents.

So, the graying of America continues. At the turn of this century, approximately 4 percent of the United States population was over sixty-five. Today, that percentage has climbed to 14 percent. More than 70 percent of people now live to the traditional retirement age of sixty-five; nearly three times as many did at the turn of the century.

Will the aging boomers seek only shuffleboard and a rocking chair in Florida? Will they be content simply to babysit their grandchildren? Or will they want more? But aging boomers, still feisty, and with a longer life span than they had anticipated when they were born, may demand more of themselves, of government, of their children. How to respond to their needs may be viewed as the most significant social challenge of the twenty-first century.

Two-thirds of the improvement in longevity has occurred during the last century. There are more senior citizens over sixty-five than there are teenagers. Most middle-aged Americans have more living grandparents than children.[17] While it is evident that women live an average of seven years longer than men, why the discrepancy exists remains unexplained. There are also racial differences life expectancy with Caucasian women living six years longer than women of African American descent. Caucasian men live about eight years longer than African American men. While the entire population of the United States has tripled since the turn of the century, the absolute number of older persons, currently thirty-three million, has increased eleven fold.[18]

Global Demographics

Among the world's richest countries—Western Europe, Japan, the United States, and Canada—the problem of an aging population has attracted increasing attention from industry and government. The aging crisis of the developed countries can also be expected to overtake the less affluent nations as well and these developing areas of the globe have fewer resources with which to cope with the "graying of the Third World."[19]

In his study of aging among less developed nations, Nicholas Eberstadt of the American Enterprise Institute contends that with the exception of the countries of sub-Saharan Africa, where even twenty years from now the median age will remain but twenty years, and in the Arab/Islamic world where total fertility rates continue to rise, "population aging is driven mainly by low birth rates rather than by long life spans."[20]

In China, between 2005 and 2025, about two-thirds of the total population will occur in the sixty-five-plus ages, or about 200 million people. Most of the elderly will have to find a means of supporting themselves inasmuch as the weak pension system makes it difficult for the Chinese government to offer much assistance. Moreover, in a country in which it is policy to discourage large families, the Chinese aged will be unable to rely on government, and likewise, will be unable to rely on family for financial support and care.

In Russia, as total population falls, median age rises so that by 2025, 20 percent of the population will be sixty-five or older. Moreover, Russia has suffered substantial deterioration in its public health system; the result that life expectancy is lower today than forty years ago. The dilemma for Russia, according to Eberstadt, is where best to put its scarce resources. "Should Russian resources be channeled to capital accumulation, or to consumption for the unproductive elderly?"[21]

In India, the demographics of an aging population are complex. In the north (Young India), the population by 2025 will remain relatively young. In the south (Old India), the population by 2025 will be aging unmistakably.[22] How will India support this elderly population with income levels lower than that of Japan and Western Europe? Aggravating matters for India's demographics is that while the population of Old India is aging, its population is educated, while Young India's relatively youthful population requires schooling. Eberstadt puts India's demographic dilemma this way, "Educated and aging, or untutored and fertile, this looks to be the contradiction and the constraint for India's development in the decades immediately ahead."[23]

The demographics of aging people in America may be examined in another way as follows:

*A 75-year-old man can expect to live eleven more years.
*A 75-year-old woman can expect to live thirteen more years.
*A 65-year-old man can expect to live seventeen more years.
*A 65-year-old woman can expect to live twenty more years.
*30 percent of women between 80 and 102 still have sex.
*By age 35, people lose more bone density than they make.
*By age 40, the waistline measurement in inches, and the risk for heart attack dramatically increases.
*By age 45, disease becomes a bigger mortality threat than accidents.

*63 percent of men between the ages of 80 to 102 still have sex.
*65 validated centenarians have lived to 110 or beyond.
*70 is the new 65, based on the health of 65-year-olds in 1973.
*74 is the average life expectancy for a boy born in 2001.
*80 is the average life expectancy for a girl born in 2001.
*85–94 is the fastest growing age group in America.
*120 is the estimated potential life span of humans, if nothing goes wrong.
*122 is the oldest, fully authenticated age to which any human has lived.[24]

Because aging populations are a universal phenomenon, academic specialties have developed to study the problems and the prospects of the aged. The process of aging is called senescence, which is not at all the same as senility. The former term is normal, senility, which may be defined as a loss of self-control and abnormal behavior, is an illness. Many elderly never become senile, while some men and women become senile relatively early on in their lives.

Gerontology is the study of the social and cultural implications of an aging society while geriatrics, a medical specialty, deals with the health issues of growing older. To this, one may add thanatology, or the study of death. Because biological age and chronological age rarely coincide, a ten-year, ten million dollar study by the MacArthur Foundation, begun in 1984, was undertaken to determine those factors that
"conspire to put one octogenarian on cross-country skis and another in a wheelchair."[25]

No one in my situation can appreciate my feeling of sadness at this parting. To this place, and the kindness of these people, I owe everything. Here I have been a quarter of a century, and have passed from a young to an old man.
—February 11, 1861. Farewell address as President-elect Abraham Lincoln left Springfield for Washington, DC. He was fifty-five years old.[26]

Chapter 2

Throw Grandma off a Cliff

From the dawn of civilization, men and women have preferred youth to age. Even moderns can feel for the aging Egyptian scribe who recognized with anguish that he was growing older:

"Old sovereign my lord! Oldness has come; old age has descended. Feebleness has arrived; dotage is here anew. The heart sleeps wearily every day. The eyes are weak, the ears are deaf, the strength is disappearing because of weariness of heart, and the mouth is silent and cannot speak. The heart is forgetful and cannot recall yesterday. The bone suffers old age. God is become evil. All taste is gone. What old age does to men is evil in every respect."[27]

Little wonder that in Egyptian hieroglyphics a bent figure leaning on a stick is the ideogram for an old person who first appeared in an inscription about 2700 BC in Babylonia." [28]

Old Age Among the Ancients

Old age has been regarded as a period of decrepitude, as an infirmity, a boring preparation for death, and so to be denied or at least postponed for as long as possible. In the first century, one person in a million reached the ripe old age of sixty. For a person to reach the extraordinary age of seventy or eighty was widely believed the result of divine intervention. Even in antiquity, some people lived more than a hundred years. Among the ancient Egyptians, "a long life is a divine reward granted to the just."[29]

"The specter of early death, sudden death, and short life still hung heavily over Renaissance society, reinforced by periodic plagues and epidemics and by frequent deaths of young friends and family members."[30]

In 1517, while negotiating with the Medici, forty-two year old Michelangelo complained that his was an old man. Michelangelo died at the age of eighty-eight. The humanist, Erasmus, at the age of forty, wrote a poem, "On the Discomforts of Old Age." Erasmus died at seventy. Dante Alighieri, who wrote the *Divine Comedy* while in his mid-thirties, died at age fifty-six.[31]

While life was often short, there have always been old people, most of who were not treated well by the societies in which they lived. In some societies the Greek historian, Herodotus, reminds us the aged were worshiped as gods, and in other societies, the aged were eaten or buried alive. Some societies fed their elderly to vicious dogs, others shouted with glee as they threw their elderly from high cliffs to be crushed upon the rocks below.[32] Early Romans disposed of their elderly by drowning them in the Tiber.

With elaborate ceremony, Samoans buried their elderly alive often with the collaboration of the victim. The Siriono nomads of the Bolivian forest unceremoniously left their elderly behind as they relentlessly sought new sources for food. The Hopi, Creek, and Crow Indians constructed special huts away from the village where they left their elderly to die. The Eskimos left them in snow banks, on ice floes, or forced them to paddle away in a kayak never to return.[33] The Fijians, on the other hand, believed a person lived eternally in death at the age at which he or she died. To assure themselves of healthy bodies for the next life, some would voluntarily kill themselves before they became decrepit.[34] What these practices demonstrate is not that the ancients were crueler to their elderly than their modern counterparts were, but that then as today, there is an ambiguity about what to do with the aged.

When an ancient society depended upon oral traditions to pass on the tribal heritage, the elderly enjoyed a respected position. When written communication became common, the usefulness of the old was in doubt. With few resources to support the senile, the elderly, with memory gone, and no longer needed to instruct the young in the traditions of the group, were killed off.

Abuse of the Elderly

In parallel manner, when the aged among us cannot quite cope with the computer, the Internet, cell phones, the iPod, the BlackBerry, or Web sites, the young appear amused at first and contemptuous before long. Impatient to receive their inheritance, the young yearn to follow the Romans and throw Grandpa or Grandma off a cliff. But, not quite able to bring that off, the young send the old to a nursing home, now euphemistically called a complete care facility. "Modernization has thus far tended to devalue old people and to reduce their status."[35]

Abuse of the elderly, although ever-present, has likewise become something of a recent discovery, as did domestic abuse and child abuse a century ago.

*An 86-year-old suffers a broken elbow after an assault by her son.
*A 73-year-old collects cans to pay for food after her daughter robbed her.
*An 85-year-old, disillusioned by her children's ongoing verbal abuse, flees home bound for nowhere in particular.[36]

Terry Fulmer, dean of the New York University College of Nursing, defines elder mistreatment "as the psychological, physical, or financial abuse, neglect, or exploitation of an older person by someone responsible for their [sic] care."[37] However, mandatory reporting of mistreatment of the elderly is getting ever-wider recognition and support.

Today forty-six states have elder mistreatment reporting laws, which require health care workers to report on those elderly who appear physically abused (bruises, fractures, etc.), or poorly nourished (malnutrition), or mentally put upon (depressed). Dr. Louis Sullivan, who served as surgeon general between 1989 and 1993, declared, "elder abuse is a health care problem not just a law-and-order issue." For many the golden years have turned to dross.

In an attempt to be more caring, we provide the elderly with marginal existence in the form of Social Security and then debate such questions as how much they need to get by and how much we ought to tax ourselves to support them. These half-hearted efforts to help make life for the elderly comfortable, cannot hide the fact that the old "are devalued, viewed as invidious stereotypes, excluded from social opportunities; and they lose roles, confront severe role ambiguity in later life, and struggle to preserve self-esteem through youthful self-images."[38]

The distinguished Canadian physician Dr. William Osler, who later became Sir William, believed that men and women over forty were essentially useless and that those over sixty should be dispatched mercifully with chloroform.[39]

Sir Peter Medawar, a distinguished physician and biologist, had this to say as recently as 1981. "In forty years time, we are to be the victims of at least a numerical tyranny of greybeards ... killing people painlessly at the age of seventy, is after all, a real kindness."[40]

In 1987, Dr. Donald Gould, a medical journalist, expressed similar sentiments when he urged that those people over seventy-five be killed off. [41]

Dr. Alexis Carrel, a brilliant French surgeon and winner of the Nobel Prize in 1912, showed little compassion for the elderly and with customary arrogance asserted, "Why should more years be added to the life of persons who are unhappy, selfish, stupid, and useless. The number of centenarians must not be augmented until we can prevent intellectual and moral decay, and also the lingering diseases of old age."[42]

These views may have been expressed with tongue in cheek, but in making them public was there more than a whiff of wishful thinking on the part of those who see in a holocaust of the aged a solution to age-old problems and problems of the aged.

The New World and the Elderly

Few countries worship youth more fervently than America. After all, America was the New World where immigrants joined the American melting pot. Newcomers were urged to reject the traditions of the Old World.

Alexis de Tocqueville, who brilliantly observed and eloquently wrote about early nineteenth century America, had this to say. "In the midst of the continual movement that agitates a democratic community, the tie that unites one generation to another is relaxed or broken; every man there readily loses all trace of the ideas of his forefathers or take no care about them."[43]

The price Americans have paid for living the American Dream has been to turn one's face to opportunity and away from aging parents. The pursuit of opportunity has encouraged Americans to establish independent homesteads early in their young adult lives, and without perhaps intending to do so, encourage their elders to shift for themselves.

In early America, men and women of advanced years were often wracked with pain. Benjamin Franklin, greatly respected at the Constitutional

Convention because of his old age, suffered from kidney stones and gout and could get by only by taking opium to relieve the pain. In a congratulatory letter to President George Washington, Franklin commented on his own health. "For my own personal ease, I should have died two years ago, but though those years have been spent in excruciating pain, I am pleased to have lived them since they brought me to see our present situation."[44]

As he lay in misery on his deathbed, his daughter tried to console him by telling him that he would live many years longer. "I hope not, he replied."[45]

David Hackett Fischer concludes, "Washington, Adams, Jefferson, almost any other prominent Americans of that era might equally serve as an example. Scarcely anyone experienced a serene old age."[46]

Such were the limitations of age among the elites of early America. How much worse it must have been for "We the People."

Because "there has never been a golden age for the old …"[47] we try to fight its coming. And so, the natural herbs and spices, the vitamin pills and lotions, the health fads and diets, the gyms and the swims, the push-ups and the nose jobs, the tummy-tucks, and the breast reductions or augmentations—thus, "Stay Young Forever" is the mantra of the day.

Grow old along with me! The best is yet to be.
The last of life, for which the first was made.
—Robert Browning[48]

CHAPTER 3

Centenarians: The Health Houdinis

The Aged Christian's Companion, first published in 1829, went through three editions by 1852 to become one of the more popular of a growing number of books on advice to the elderly. Written by John Stanford, a Christian minister, *The Aged Christian's Companion* sought to advise old people how to live better lives. Consistent with his theological training, Stanford defined old age as the biblical threescore years and ten. "As in the natural world," Stanford declared, "so it is in the human creation. Times, seasons, and periods of existence are fixed, and they cannot pass their bounds."[49] Stanford observed that most people seek to live to a ripe old age, but "death often blasts the prospect!"[50] As one would expect of a minister, John Stanford urged his elderly readers to "find grace in the eyes of the Lord,"[51] to assure a contented old age.

Finding Grace in Aging

During the late nineteenth and early twentieth centuries, elderly women and men were admonished that like Jacob, who wrestled with God, good and bad old age likewise wrestle with each other. A good old age is one "of virtue, health, self-reliance, natural death, and salvation." A bad old age was one of sin, disease, dependency, premature death, and damnation."[52] A virtuous youth assured the former, a self-indulgent one led to the latter.

In an 1862 article in *Atlantic Monthly*, Ralph Waldo Emerson took a more secular view as he wrote of the advantages, rather than of the infirmities, associated with growing old. "America is a country of young men, and too full of work hitherto for leisure and tranquility; yet, we have robust centenarians and examples of dignity and wisdom." Emerson quotes President John Adams, who at ninety was, in 1825, still alive to witness with pride the election to the presidency of his son, John Quincy Adams. Adams Senior goes on to explain to his visitors that he and presidents Washington, Jefferson, Madison, and Monroe were all about fifty-eight and considered well on in years when they assumed their high offices.[53]

By the twentieth century, however, it was apparent that virtue alone was insufficient to assure a serene old age. Advances in science, community medicine, medical care, and better knowledge of diet and exercise, and health practices urged upon us have helped make a good, not a debilitated old age possible. Indeed, some scientists have suggested that we can readily live to be a hundred and others, dare ask, "Can we live forever?"

The Legends of Age

During the Jewish Golden Age, patriarchs allegedly lived hundreds of years, including Adam (930 years), Seth (912 years), Enosh (905 years), Kenan (910 years) Mahalalel (895 years), Jared (962 years), Enoch (365 years), Methuselah (969 years), Lamech (777 years), and Noah (950 years). Moses was eighty years old when God called upon him to free the Jews from slavery in Egypt. Apparently, God did not think him too old for so formidable a task. Moses led his people in the desert for forty years and died at the age of one hundred and twenty without seeing the Promised Land. Whether literally or metaphorically interpreted, the once seemingly impossible goal of leading a full and rewarding life to the ripe old age of one hundred and twenty becomes an increasingly realistic age for the life span of humankind. "May you live to a hundred and twenty," an oft-repeated birthday toast may become more than just a figure of speech during the decades to come. If some people could live to be more than a hundred years old before, why not now?

Legends linger of people in ancient and modern times who have lived and continue to live for many years beyond the biblical "threescore and ten." Greek geographer Pliny in his *Natural History* asserts that in remote areas of the then world, men and women lived to an extreme old age, "until sated with life and luxury, they leapt into the sea."[54] Do such Shangri-la's populated by those enjoying good health and immortality as depicted in James Hilton's 1933 novel *Lost Horizon*, still exist?

Alexander Leaf, MD thought so. In an article for *National Geographic*, he wrote, "There are places in the world where people are alleged to live much longer and remain more vigorous in old age than in most modern societies."[55] On a two-year sabbatical, Leaf visited remote and mountainous villages of Vilcabamba, Ecuador, in the Andes; Hunza in Pakistani controlled Kashmir, and Abkhazia in the former Georgian Soviet Socialist Republic. "He sought to verify age through baptismal records, testimony by friends, and inconsistencies in the memories of the centenarians." Dr. Leaf was convinced that while their ages may not be exact, those interviewed were clearly well over a hundred years old, and that in each of these regions could be found centenarians who were, as a proportion of the population, considerably greater than in the United States. Unless we are sure such enclaves of extreme longevity do not or cannot exist, are we not obligated to try to find them?

Another article in *National Geographic*, "The Secrets of a Long Life," Dan Buettner sought to find out why Okinawans, Sardinians, and Seventh Day Adventists, appeared to live a long time and escape many of the diseases that often terminate life. The Sardinians, for example, quaff red wine, albeit in moderation, actively work their farms, eat fruits and vegetables, and have a warm and engaging family life. Okinawans likewise have a wide network of friendships but eat moderately. The Seventh Day Adventists in Loma Linda, California, are among "longevity all-stars." Residents of these three places enjoy more healthy years of life and produce a high rate of centenarians.[56] Those geographic areas in which people regularly live healthful lives even to one hundred, Buettner called these regions "Blue Zones." Among them, Buettner subsequently found a group of centenarians who live on the Nicoya Peninsula in Costa Rica.

Centenarians are often also contrariness. That is, they eat small portions of simple food rather than large portions of rich food. They do not smoke and many drink red wine. Mostly, however, these people are not driven to succeed, maintain self-confidence in the future, and for the most part, have freed themselves from anxiety. They generally have strong family ties and a lively social life. Many have felt the sting of tragedy as when sickness or accidents bring premature death to children and loved ones, yet through it all maintained a sunny outlook and unshakable faith in the future and in their good fortune.[57]

The Health Houdinis: the Centenarians

In 1949, six, one-hundred-year-olds were among the Union Soldiers at the reunion of the Grand Army of the Republic. With an average life span of 103, these sturdy soldiers traveled some distance from their homes and were healthy enough to participate in the events of the reunion with humor and gusto. "These men were more than a generation older than the average life expectancy of the general population in 1949 and almost two generations older than the average life expectancy when they were born."[58]

There are about 40,000 centenarians in the United States, or a little more than 1 centenarian per 10,000 in the population; 85 percent are women, 15 percent are men. American centenarians are as diverse as the American population. Some still work at paid jobs; others are volunteers helping their neighbor or their community. Some are heavy drinkers others are teetotalers. Some were born slaves; some seemingly enjoyed all the breaks life had to offer. Others experienced great tragedies to themselves or more likely to their families. Many, perhaps most centenarians, are healthy and still have at least some of their own teeth. Few are obese. Many can read the newspaper without eyeglasses, and some have never seen a doctor while others have been sick, bedridden, or confined to a wheelchair for many years. Some centenarians still ride horses, others still drive cars, and occasionally, one can find a centenarian who drives a motorcycle. As a group, centenarians make and keep friends and keep romance alive. These close relationships help contribute to their longevity.[59]

According to Dr. Thomas Perls of the Harvard Medical School and Boston's Beth Israel Deaconess Hospital, the most important factor determining longevity is the genes with which our forebears endowed us. With data drawn from interviews with centenarians, evaluation of their mental and physical health, and blood samples, Perls directed the New England Centenarian Study and came to the unsurprising conclusion: "They've (centenarians) got genes that are facilitating the ability to age very slowly and they lack genes associated with aging diseases. It is definitely the two together." [60]

In 1988, Vianno Kannisto, a Finnish demographer, reported on a study based on persons who reached one hundred years of age in Australia, Austria, England, Wales, West Germany, Finland, France, Iceland, Italy, the Netherlands, New Zealand (excluding the Maoris people), Norway, Sweden, and Switzerland—about 280 million people in countries where good birth and death records are maintained. What did he find? He found that "Only 67,286 persons out of this very large population reached 100. About half that number reached 101, a quarter lived to 102. At 108 years of age, 219 females and twenty-two males survived. At 109, seventy-four females and six males, and thereafter only females: twenty-seven aged 110, three aged 111, and one aged 112."[61]

Since becoming a centenarian remains rare, it is little wonder that in some countries those who have lived for a hundred years are granted special recognition. Japan, which in 2005 had the largest number of centenarians in the world, honors them with a silver cup and a certificate from the prime minister. In Japan, September 15 is National Respect for the Aged Day and a national holiday. In the United Kingdom, the queen sends greetings to each person who turns one hundred and sends good wishes each year to those who have achieved the age of 105. In the United States, there is a growing tradition that upon reaching one hundred years, American centenarians receive a congratulatory letter from the president.

While living to be a hundred is rare, it is also no piece of cake. As one centenarian explained, "Growing old is a terribly hard thing ... a person loses all his friends. They die and you are left alone"[62] While living to one hundred is not for sissies, medical progress has made centenarians the most rapidly growing age-cohort in the world.

Traditionally, the oldest people seem to come from Japan, Norway, or Sweden. Today, of the thirty oldest people in the world, half now come from the United States. Of this group, the three oldest are American women. While most centenarians are overwhelmingly women, men seem to be catching up. In the United States, according to a study by the Harvard Medical School, the number of centenarians doubled in the 1980s and did so again in the 1990s. The total now exceeds 70,000. By 2050, more than 800,000 Americans could celebrate the century mark.[63] Female centenarians outnumber males nine to one. While only 15 percent of centenarians are men, those men who reach the century mark seem healthier than the women are. "The chances of reaching 100 have approximately doubled every 10 years in the recent past, with the chances for women being about four times those for men."[64]

The Super-Centenarians

The Gerontology Research Group (GRG) estimates that there are also super-centenarians among us who can document that they are 110 or older. According to the GRG, despite fraudulent claims to the contrary, the current age barrier is 114, and no one has breached "the invisible barrier of 115."[65] Yet, some claims to this effect have been made. In August of 1997, Jeanne Calment of France apparently lived to age 122 "the longest well-documented life on record."[66]

The Guinness Book of Records lists Shigechiyo Izumi (1865–1986) as the oldest male it has ever recognized. Yet the Guinness editors remain very cautious about those it lists who have lived a very long time. "No single subject is more obscured by vanity, deceit, falsehood, and deliberate fraud than the extremes of human longevity."[67] Are these records of significance? Or are they merely aberrant records that defy classification?

James M. Carey believes they are important milestones not only in the lives of the lucky few who attain them, but for the rest of us and for those who follow. "Changes in record ages," he argues, "are important, because they are the harbingers of the future and the extreme manifestation of improved health. It is virtually certain that persons living far beyond the current age record of 122 will begin to appear in the twenty-first century, ages of 130 years will likely be reached before the end of the century."[68]

Since the United States has become the home of most centenarians, Appendix B offers biographical vignettes of some Americans who have reached these very considerable milestones. While the super-centenarians cannot explain with any precision why they lived so long, are we about to break through the 114 year-old age barrier? Would it be desirable to do so? How might we achieve it? Were we to do so, could we then say that humankind has the potential for living forever?

The creed of the street is Old Age is not disgraceful, but immensely disadvantageous.[69]
—Ralph Waldo Emerson

Chapter 4

Gullible's Travels

According to one interpretation of the Bible, Jewish scholars explain that during the age of the great patriarchs, people grew old but retained their youthful vitality. Death occurred suddenly and painlessly on a predetermined day when the soul departed through a sneeze. Hence, the age-old expressions, "Bless you," or *Gesundheit*! when a person sneezes.[70] Saint Augustine attributed Adam and Eve's ever-youthful vitality to the wonderful grace of God. Expelled from paradise, however, they became subject to the process of illness and aging that has been the fate of humankind ever since.

The study of aging has often attracted crackpots, cranks, and medical quacks who appeal to a willing band of patients who are looking for a quick and easy restoration of youthful vigor, and for good measure, perhaps a tincture of immortality as well. What follows are vignettes of respectable, if often-misguided seers who were sure that the prolongation of life, and even immortality, were at hand. Olshansky and Carnes warn, however, like gamblers in a casino, "The likelihood of striking the longevity jackpot is similarly small."[71]

Juan Ponce de Leon (1460–1521)

Juan Ponce de Leon, a soldier in the royal household of the King of Aragon, accompanied Christopher Columbus on his second journey to the New World in 1493. Ponce de Leon, ordinarily a cautious adventurer, thought there may be more reality than fable in the stories that in the lands now known as Bimini and in the Bahamas, there exists a perennial spring of water from which those who drink from it makes old men young again.

When reports of a fountain of youth surfaced, a question stirred considerable interest and discussion: was the search futile or did it have a serious intent. In the early years of the sixteenth century, Peter Martyr d'Anghiera, a member of the Council of the Indies and one who was close to the Spanish conquistadores, gave the pope his opinion that the search should be seriously undertaken. Thus, he wrote to the Pope Leo X:

> "Let not Your Holiness believe this to be a hasty or foolish opinion, for the story has been most seriously told to all the court, and made such an impression that the entire populace, and even people superior by birth and influence, accepted it as a proven fact."[72]

With the blessing of the pope and king, Ponce de Leon went off on a serious quest for the fountain of youth.

For Ponce and his men, the fountain's water was the very *Viagra* of life, a sure cure for erectile dysfunction and fading virility. Little wonder that Ponce de Leon, at age fifty-five, led a systematic search for this remarkable body of water. For fifteen years, as Governor of Puerto Rico, he sought the elusive fountain of youth "whose water rejuvenated the aged." An Indian arrow struck and killed Ponce de Leon. He died at the age of sixty-one.

The idea that cavorting in waters may restore youth and vitality may be traced to Hindu legend and to the Old Testament. Even Aristotle suggested that vitality could be restored to the human body by immersion in magical streams. Alexander the Great, Aristotle's most prominent pupil, likewise accepted the theory of a restorative body of water and searched for yet another fountain of youth.

In our day, the alleged efficacy of taking the waters either by bathing in foul-smelling sulfurous springs or by drinking waters from special spas, lingers in now aging resorts such as Carlsbad in Germany, Yalta in Russia, Vichy in France, Bath in England, and even in Saratoga Springs, New York.

The fountain of youth was never found, but in the long search, Ponce de Leon conquered Florida and encouraged the quest for immortality. However, no trace of a medicinal river with the power to make the old young was ever found. Nevertheless, just as we are not likely able to live forever, the search for immortality may lead us to scientific highways and byways, and in the process, conquer the diseases of age, and therefore, perhaps make our existing years, however short or long, productive and rewarding ones.

Luigi Cornaro (1464–1566)

A man of the Italian Renaissance, and indeed a Renaissance man, Cornaro came from a well-connected Italian family who had significant business and political interests in Padua, where he was born and in Venice where he lived. He was a patron of the arts, a sponsor of significant architectural projects, and a friend of Raphael. However, by the age of forty, having lived *La Dolce Vita* (The Good Life), he had dissipated his health, and under doctors' orders, altered his lifestyle and his eating and drinking habits. Unlike those of us who make New Year's resolutions they do not or cannot keep, Cornaro was determined to live long by foregoing his delight in rich food, good wine, and sensual pleasure.

So successful was he at living *La Vita Sobria* (The Temperate Life) that by eighty-three, he was requested to put his health formula into writing so that others may follow it. By the time of his death, he had written four volumes on health as a guide to others who sought to regain their health and add to their longevity. Cornaro was never in doubt that a long, healthy life, not a short sickly one, was essentially, what God desired. "Live, live, that you may become better servants of God."[73]

Moreover, Cornaro believed that old age need not be grim but could be a happy time. He wrote, "I have an ardent desire that every man should strive to attain my age, in order that he may enjoy the most beautiful period of life."[74] He boasted that despite advanced age, he could mount his horse unaided, climb stairs, or traverse a steep hill with ease. He maintained his teeth well, he fell asleep easily, and he awoke refreshed, rested, and ready to entertain his eleven grandchildren.

While he believed that God did not intend human beings to mortify the flesh, he deliberately limited his diet. His food consisted of small portions of bread, meat, broth with egg, and new wine. The main message for dieters was to eat the foods they liked in moderation. Meager though the diet sounds, Cornaro insisted, "I always eat with relish, and I feel when I leave the table, that I must sing."[75]

Cornaro was wrong, of course, when he asserted, "A man can have no better doctor than himself." Yet, it is not surprising that he should believe this, as did many other prominent people. Cornaro was living in a prescientific age, when "it was still possible to envision the control of disease by means of very simple measures."[76] He could not have been aware of the role advances in medicine and public health and the influence it would have on human longevity. The simplicity of Cornaro's approach to longevity is with us in the nostrums and quack medicines readily available, in the refusal of some to accept the impact of evolution, in the obstinacy with which some turn their backs on the longevity that might be achieved through advanced research on embryonic stem cells and genetic modification. Yet, Cornaro lived in good health to the age of ninety-eight. He must have done something right!

Elie Metchnikoff (1845–1916)

The Russian born bacteriologist, Elie Metchnikoff, who had won a Nobel Prize for his work on immunity by the age of twenty, had already published a number of scientific articles, which withstood critical review of jealous scientists. With Metchnikoff, the search for a fountain of youth shifted from a search for a mythical spring, whose waters could bring health and well-being, to a search in the laboratory for the underlying causes of aging. In 1888, Metchnikoff was appointed to the Pasteur Institute where, at the nexus where zoology, biology, and medicine intersect, he worked for the rest of his life.

Metchnikoff, however, uncritically accepted the claims that Kentigern, founder of the cathedral at Glasgow, had died at the age of 185 in the year 600, and that Pierre Zortray, a Hungarian farmer, likewise died at 185 in 1724, and that Thomas Parr died at the age of 152 and was buried in Westminster Abbey in 1635.[77] Metchnikoff could not be convinced that some skepticism was appropriate. Before long, living to a mere 140 would be the rule rather than the exception. So he thought.

In his book, *The Prolongation of Life* (1907) Metchnikoff wrote, "When we have reduced or abolished the causes of precious senility as intemperance and disease, it will no longer be necessary to give pensions at the age of sixty or seventy years. The cost of supporting the old, instead of increasing, will diminish progressively."[78] He assumed that with a longer life span, men and women would work for many years and would draw less on such pensions that may be available to them. Many were not so sure.

In France, where Metchnikoff worked, many legislators believed that the nation could not support men and women over seventy without bankrupting the nation. Besides, they asserted, what if they were not only a burden on society, but also a burden to themselves. What if they lived alone and independently and required medical care? Metchnikoff recognized that if he developed his theories to their conclusion, the very nature of old age would have to be researched. Did old age always involve senility? Was old age a sickness? How best to live in one's later years? He sought to demonstrate through strict scientific models that aging could be thwarted, perhaps even, indefinitely.

Metchnikoff's views became popular in the mass media and widely circulated in the United States. His theories of diet led to fads including such alleged age enhancement foods as soured milk, kefir, sauerkraut, salted cucumbers, yogurt, all of which would help the body reduce the poisonous bacteria most diets encouraged. A diet heavy with nourishing, if often hard to ingest foods, would help destroy bacteria, which attacked the nervous system, liver, hair, and kidneys. Thus, would the toxic substances of traditional diet be destroyed and life prolonged? He wrote "fermentation and putrefaction harmful to the organism" could be conquered.[79]

In 1902, Metchnikoff encouraged the production of sour milk by a Paris company thus "sparking an international sour milk craze."[80] Although Metchnikoff backed away from his more sensational claims, some serious scientists continued to insist that sour milk was the secret to living forever!

Charles-Edouard Brown-Séquard (1817–1894)

On June 1, 1889, a group of distinguished biologists gathered in Paris to hear a lecture from six-foot-four, seventy-two year old Charles-Edouard Brown-Séquard, a prominent physiologist, who had held a series of distinguished positions in Europe and America, which included teaching at Harvard and at the National Hospital for the Paralyzed and Epileptic in London. He was also professor of comparative and experimental pathology at the Faculty of Medicine in Paris. He had been born on the British owned island of Mauritius to a French mother and an Irish-American sea captain father. Along with his distinguished professional career, he married thrice including a woman much younger than he was.

Professor Brown-Séquard, imposing in appearance, towered over his audience of fellow scientists who eagerly awaited the latest dramatic breakthrough they had come to expect from so formidable a scientist. "Brown-Séquard described for his colleagues the lassitude and physical impotence that had overtaken him." He then went on to describe how he had cut the testicle off a young dog, mashed the pieces with juice and water, and then injected the solution into his leg. He described how he repeated the process using guinea pig testicles, and once again injected the solution into his leg. After a third injection, he described that a remarkable transformation had taken place. "He felt like a young man again." Then he boasted, "Today I was able to pay a visit to my young wife."[81] He said he felt like a teenager, but was he boasting like one as well?

Whether boasting or not, the Paris newspaper, *Le Matin*, sought to raise funds to replicate Brown-Séquard's work to bring welcome relief to other frustrated men. Brown Séquard, author Osborn Segerberg wrote, contrived a contraption of belts, pulleys, and bladders that yielded a liquid that treated the impotence of elderly men.[82] It is unlikely that Brown-Séquard long enjoyed his popular success. His young wife deserted him, he died five years after his famous lecture, and his reputation as a scientist tattered.

Linus Pauling (1901–1994)

Among his many awards, Linus Pauling received the Nobel Prize in Chemistry in 1954 and received the Nobel Peace Prize in 1963. The only previous winner of two Nobel prizes was Marie Curie who won one prize in 1903 and another in 1911.

In 1974, President Gerald Ford awarded Dr. Linus Pauling the National Medal of Science and in 1989, the National Science Board recognized his contributions to science, technology, and society with the Vannevar Bush Award. Some fifty colleges and universities awarded him honorary degrees for his work for peace. In addition to the Nobel Peace Prize, he was awarded the Gandhi and Lenin peace prizes and the Albert Schweitzer

Peace medal. His wife of many years, Ava Helen Pauling, died in 1981. He always felt she deserved to share the Nobel Peace Prize with him.

When Linus Pauling died, he was regarded as one of the world's greatest and most visible scientists and was recognizable to newspaper reading and television viewing audiences. His virtuosity as a chemist led him into many humanitarian efforts, especially as an anti-war activist as well as an opponent of aboveground testing of nuclear weapons. He was among the first to identify that statistically, smoking shortened the average life span by as much as eight years.

His interest in the factors of longevity waxed and waned throughout his career, but in his later years, he became especially concerned with vitamins and the biochemistry of nutrition. He founded a new field that he called "orthomolecular medicine" (the right molecules in the right concentration). This approach led to his conviction that large doses of Vitamin C would not only cure the common cold, but would become the treatment of choice for life-threatening illnesses including the flu, cancer, and AIDS. His thinking went so far as to urge that vitamins in mega doses could slow the aging process.

His 1970 book, *Vitamin C and the Common Cold* sold widely to become something of a medical fad, not unlike the popularity of sour milk that Elie Metchnikoff endorsed. For all his accomplishments, Linus Pauling is best known for his advocacy of Vitamin C, the least laudable aspect of an otherwise illustrious career.

Although orthomolecular medicine was widely viewed by established physicians as medical quackery, this did not deter Linus Pauling from establishing an Institute of Science and Medicine to conduct research in slowing the aging process and relieving attendant suffering that aging often brings in its wake. His assertion that mega doses of Vitamin C can improve general health was never proven. A cancer study at the Mayo Clinic that involved 367 patients, failed to demonstrate that mega doses of Vitamin C helped patients with advanced cancer. Nor was Linus Pauling able to help himself. Despite treating himself with regular doses of Vitamin C, he died at the age of ninety-three of prostate cancer.

In the maze of medical reports, women and men, especially the elderly, search medical avenues in hopes of finding a quick, easy, inexpensive, and painless formula for restoring youthful vitality. And, as we have seen, prominent scientists Metchnikoff, Séquard, and Pauling were not exempt from the tantalizing vision that if death could be conquered, the good life may be enjoyed forever.

The Immortalists

In 2007, David M. Friedman wrote a dual biography of a most unlikely pair. One individual was Dr. Alexis Carrel, a fifty-seven-year-old short, chubby physician who in 1912 at the age of thirty-nine became the first man of American medicine to win the Nobel Prize for his work in suturing blood vessels. The other individual was Charles A. Lindbergh, "The Lone Eagle," who in 1927 flew his single-engine plane, the *Spirit of St. Louis*, across the Atlantic. He landed at Le Bourget Airport in Paris where he became an instant hero to the French and to the world. Lindbergh, tall and spare in build, Dr. Carrel had to stand on his toes to look his visitor in the eye to greet him properly.

What the two men had in common was a deep interest in what machines might do to improve the human condition, perhaps even to achieve immortality. Besides being an aircraft pilot extraordinaire, Lindbergh was also a grease monkey. He loved to see how things worked. Carrel, for his part was a blood monkey, and with extraordinarily deft hands could repair broken or aged valves in the human body. As Lindbergh saw it, if a damaged valve on his single-engine airplane could be repaired, why not the diseased valve in the heart of his sister-in-law.

In his search for a solution, Lindbergh had come to the right man inasmuch as Carrel had performed the first coronary bypass operation on a dog in 1910. Lindbergh must have been disappointed when the daring but cautious surgeon believed that the state-of-the-art in heart surgery made what Lindbergh proposed impossible. The problem had not been shown how to keep the heart alive outside the body so that repairs could be made, and the still beating heart returned to its appropriate cavity in the chest. In his tour of Carrel's generously proportioned laboratory in Rockefeller University in New York City, Lindbergh observed that Carrel had kept some animal tissues living, some for years, outside the human body. Could the heart, likewise be kept alive outside the human body?

Together, the two men developed a perfusion pump, which made it possible for organs that had been removed from a patient, could be kept alive by circulating fluid around the detached organs. While this invention had its shortcomings, it became widely regarded in the medical community as the first artificial heart, and a significant step toward organ transplant surgery.

A common project of uncommon significance, however, was not enough to make the two men bond. What made their collaboration possible was a shared view of life. Both men sought to square the scientific with the divine. Sadly, however, both men having achieved success early in life began to feel that they were divine and placed their scientific and technological acumen at the service of the eugenics movement. Single-mindedly, they sought to breed a race of heroes, a master race. They viewed themselves as arbiters of what a master race could be like.

To Lindbergh, entering the laboratories of Dr. Carrel was nothing less awe-inspiring than entering a cathedral. This concept was strengthened in that Carrel and his medical associates wore black not white. They wore black hoods reminiscent of the cowls worn by monks. The black painted walls of the operating rooms projected the environment of a monastery not a hospital. The monastic rather than the scientific frame of reference prevailed. Young Lindbergh was impressed.

In his 1935 book, *Man the Unknown*, Carrel asserted, "Man is the hardiest of animals, and the white race, builders of our civilization, is the hardiest of all races."[83] Because "eugenics, may exercise a great influence on the destiny of the civilized races," Carrel proposed that the feebleminded, the insane, be prevented from propagating. He proposed that a medical examination be required of people about to be married. It was but a short step from this to Hitler's gas chambers. Lindbergh, the all-American hero should have known better, but the sophisticated Parisian and New Yorker held the farm boy from the Midwest in thrall. Thus, Carrel offered his services to Marshall Petain, head of the French puppet regime, while Lindbergh became a leader of the semi-fascist America First Committee.

Gullible's Travels

For every prominent physician or scientist who squanders his or her reputation on what turns out to be a medical fraud, there are equally distinguished figures of worldwide fame, who squander their wealth in a generally vain attempt to hold onto life and to power a little longer. For many, one life does not seem nearly long enough to achieve all one wishes to achieve.

Ana Aslan, a Romanian physician, promoted Gerovital and H3 and boasted that among those taking her preparation included French President Charles de Gaulle, German President Konrad Adenauer, the novelist and playwright, William Somerset Maugham, Mao Tse-tung, Pablo Picasso, Marlene Dietrich, Aristotle Onassis, and John F. Kennedy. Although Dr. Aslan died at the age of ninety-six, no breakthrough was ever achieved by the distinguished or by the not so distinguished men and women who tried "the cure."

Dr. Paul Niehan promoted something called "cellular therapy." His cure consisted of injecting patients with fresh cells from unborn lambs. Among the celebrities treated at Dr. Niehan's sumptuous clinic called *La Prairie* in Montreux, Switzerland, were, Konrad Adenauer, Charlie Chaplin, Pope Pius XII, the Duke of Windsor, Bernard Baruch, Winston Churchill, and Christian Dior.

Making a fortune, more than developing a reputation as a scientist, has motivated some to encourage others to believe that virility and vitality could easily be achieved and retained forever. The Russian physician, Dr. Serge Voronoff, made his fortune by transplanting ape testicles into old men as a form of rejuvenation therapy. The poet, William Butler Yeats, thought such a transplant would help him court the actress Maud Gonne. The transplant and the courtship both failed. John Romulus Brinkley transplanted young goats' testicles into old men. He made money and ran for governor of Kansas, but his patients failed in their quest for more rewarding sex. For those who succumb to the lure of prolongevity, if not immortality, by seeking the cure first at one elegant spa and then at another, or by seeking an exotic if unproven transplant of sex organs, Dr. Roy L. Walford has described the search as "Gullible's Travels."[84]

What harm would there be if I were to live, if not forever, then for a few more years? Moreover, yesterday, the moon, today the pill, tomorrow the morning-after pill, the day after tomorrow, immortality? Those who inveigh against the futile search for immortality are betting against the odds. Those who insist that immortality is neither desirable nor possible will not prevail. The holy grail of perpetual life may not be found, we may not return to the Garden of Eden, we may not eat once again from the tree of life, yet who knows what serendipity might yield as the search for immortality continues.

Gullible or not, the siren song of our time remains the search for life everlasting. Immortality is a lure that will not easily dissipate. No one wants to die. Cheating death may be an idea whose time has come.

The Reader will easily believe that from what I had heard and seen my keen Appetite for Perpetuity of Life was much abated.[85]
—Jonathan Swift

Chapter 5

Do You Want to Live Forever?

In *Gulliver's Travels* (1726), Jonathan Swift, perhaps the Beatrice Potter of his day describes the immortal struldbrugs as the least fortunate of people. Although they live forever, their lives become little more than a long-drawn-out death. As the oldest in their community, they ought to be honored and revered by others, but instead they are shunted to one side as they demonstrate the mental quirks and physical fragilities of age without redeeming qualities. As Jonathan Swift describes them:

"They were the most horrifying Sight I ever beheld, and the Women more horrible than the Men. Besides the usual Deformities in extreme old age, they acquired an additional Ghastliness in Proportion to their Number of Years, which is not to be described; and among half a Dozen I soon distinguished which was the eldest, although there were not above a Century or two between them.[86]

In view of Swift's frightening description of what happens to people who live forever, perhaps one should never really want to do so.

Reasons Not to Want to Live Forever

1. Immortality is a futile goal. Why pursue it then?
2. If living forever does not include maintaining youthful vigor and zest for life, is it worth achieving?
3. Since living forever may involve chronic infirmity, why make it a goal?

4. Living forever implies the loss of family and friends; loneliness is sure to be the lot of the extremely old.
5. If living forever means living as old as the dinosaurs, one may not be able to communicate with other beings because the very language one uses would change. In short, in living forever do we lose our humanity?
6. Have you ever experienced the torment of tossing and turning, and otherwise being unable to sleep? If immortality may be viewed as something like that, what torment immortality may be.
7. If living forever means that everything that can be experienced has been experienced, might not living forever be boring making living essentially killing time. Does eternal life inevitably also mean eternal punishment?
8. If no one dies, the young never grow up. They never inherit from their parents.
9. Perhaps it is the duty of the elderly to die to give youth a chance.
10. Immortals use more than their fair share of the resources of a finite planet and so deprives others of the environment necessary to sustain life and vigor.
11. Since initially at least, treatments and medicines needed to thwart death are likely to be expensive, is justice served if only the affluent can afford them?
12. Is death the "mother of beauty" as some poets asserts? What they mean is that beauty, as in the case of fresh flowers, does not last. And, the beauty of a Picasso or a Rembrandt is derived in part from the fact that those who made them are now dead. They are all we have and so they are beautiful.
13. If it were possible to live forever, how would we define happiness? What would make us happy? How would we define courage, and how would we be courageous? "Call no man happy until he is dead," so said the Athenian philosopher Solon. Is he right? [87]
14. If we lived forever, who would we be? Brain cells store our memories, and if new brain cells do not remember what took place before, how would we know whom we are? What we were? And, perhaps, how would we plan for the future?
13. If men and women were immortal, would heroism be possible? Do we not ennoble others and ourselves if, precisely because the years of life are numbered, we give some of them to a worthy cause, to a heroic deed, even to give up our life for principle? "Immortals cannot be noble."[88]

By wishing, at least sometimes, to live forever are we doomed to become struldbrugs? Not necessarily. In fact, whether it is desirable or not, immortality may be thrust upon us. That is, even a slight addition to the life expectancy will produce a multiplier effect so that the traditional threescore and ten years of a person's life will be a quaint number to contemplate, but by no means, a realistic assessment of what humankind can expect. "Immortality, not merely some reasonable extension of life span, is the great expectation."[89] As the Apostle Paul asserted, "The last enemy that shall be destroyed is death."

Reasons for Wishing to Live Forever

1. Those who are living the good life will want to live forever among family, lovers, and friends who are also enjoying the good life and wish to live forever.
2. Immortality helps us assess whether the plans we have nurtured in our youth will mature in the future for ourselves, our children, our grandchildren, and perhaps for the well-being of humankind. What octogenarian does not want to live long enough to dance at a granddaughter's or even a great-granddaughter's wedding?
3. Immortality may be perceived as a reward for living a virtuous life.
4. If one lives long enough, Cornaro (see pp. 00) asserted, one may have the blessing of a "natural death," rather than one marked by pain and suffering.
5. Likewise, a long life is one during which one may blossom creatively in later years when such opportunities were denied in the daily grind of making a living. After all, Grandma Moses did her best work as an "old lady," George Bernard Shaw was writing superior scripts in his seventies, and Boris Pasternak finished writing *Doctor Zhivago* when he was sixty-five. Perhaps there is some greatness in all of us, which a long life would bless.
6. If hope springs eternal, why not hope for immortality and in the process learn to cope with the social problems that may develop. How do we know how to overcome those problems if we have not experienced them?
7. If we do not try to triumph over disease and death, its constant companion, how do we know that we cannot, should not, and will not eventually do so?
8. If we are destined to lose the search for immortality, achieving the next

best thing, prolonged youthfulness and increased longevity, may not be at all bad.
9. Immortality makes it possible for women and men to anticipate more than one career, to postpone or even avoid senility. Rather than add to the social cost, men and women who live productive lives may continue to work, and thereby become less of a social burden on a nation's economy.
10. Immortality does not mean more of the same. It means different, richer, more varied as well. Life is not merely "one damn thing after another," but represents a change in life's texture and in the nature of what it means to be human. It will take surely more than one life to discover, colonize, and develop distant stars once it becomes clear that we have the capacity to do so and that it is the right thing to do. The eminent British physicist, Stephen Hawking, suggested that it may be time for humankind to consider how best to colonize other planets.
11. Does death really ennoble life, or does life ennoble life?
12. If we do not achieve immortality, in the end of days who will be there to report it? How will it get into *The New York Times* or *Washington Post*, or who will put it on the Internet? How will we know the end of the world has come?

To Alexander the Great's query, "How long is it good for man to live?" the Jain holy man replied, "As long as he does not prefer death to life."[90]

That we will grow old and die is seen as inevitable, even desirable. But what if the advances of science make it possible to interfere with what has been nature's mandate, dare we do so? Should we even try? Up to now, we have made immortality the province of God. But might eternal youth soon be in the hands of a friendly pharmacist? What if, scientifically, we can live forever, but the circumstances of the human condition make it impossible to take advantage of what science makes possible. What then for man?

Pornography or Science?

For too long, respectable researchers have viewed studies in immortality as scientific pornography, which they cannot define, yet feel they can recognize when they see it. But is it pornographic to anticipate that when computer engineers work together with surgeons it may be possible in the next twenty-five years or so to improve the disfigured face of an accident victim, or replace a diseased pancreas of a diabetic? Is it pornographic to anticipate that within the lifetimes of our children or our grandchildren that "we'll be able to check ourselves in for an overhaul late in life? Our new organs will be manufactured the way Ford makes crankshafts."[91]

Dr. Michael Rose is known for his research in manipulating the longevity of fruit flies. The Methuselah flies he developed had longer life spans than those that were not part of the experiment. He then returned to the laboratory and shortened the life of fruit flies, thereby demonstrating that at least in fruit flies, life span can be lengthened and shortened. If this can be done in a laboratory with fruit flies, why can't it be done with ordinary women and men? Indeed, Dr. Rose believes that, as with fruit flies, the evolutionary process can be accelerated, something similar can and will eventually be accomplished for human beings. In response to the query, "What will it take to increase human life span from the present level?" Dr. Rose replied,

"There is not going to be one magic bullet where you can take one pill or manipulate one gene and get to live to 500. But you could take a first step and then another so that in 50 years time, people take 50 or 60 pills and they live to be 200.

"Leaving aside F.D.A approval, it looks like we are about 5 to 10 years away from therapies that would add years to our present life span. For now, pharmaceuticals will be the primary anti-aging therapy.

"Eventually, we will be able to culture replacement organs from our own cells and repair damage using nanotech machines. All of this will increase life span."[92]

Why get a body tattoo or a nose ring when you may be able to express yourself by changing your bodily silhouette to match your mood. Maybe even add a sex organ or two to have more fun! If you want to look like Nicole Kidman or Bill Clinton, why not?

William Heseltine, founder of Human Genome Sciences, a two billion dollar drug company working to achieve what he calls "practical immortality," asserts, "The same convergence of stem cell cellular immortality, and genetic-engineering technologies ... will enable us to re-machine ourselves with youthful organs ..."[93] He continues, "All the body can be restructured. You can look like you want, have the color you want ... No more rock-hard-boob syndrome, but nice, soft, fleshy boobs."[94] Is Mr. Heseltine a seer or a salesperson or both?

Through genetic manipulation or stem cell research, a couple may have a made-to-order baby, by Yves Saint Laurent perhaps, who will grow up to be a Wonder Woman or a Superman, not unlike the science fiction comics of yesteryear.

At Natick Labs in Massachusetts, army engineers are developing a suit that "may ultimately allow soldiers to leap tall buildings with a single bound."[95] However, if, as many scientists anticipate, we may do many if not all of these things before the end of the twenty-first century, can we increase longevity beyond say 120 years, which appears to be the current outside limit? If some lobsters and some whales can live for two hundred or more years, why can't you and I?

John Wilmouth, a demographer from the University of California, seems to think so. "There is no scientific basis on which to estimate a fixed upper limit. Whether 115 or 120 years, it is a legend created by scientists who are quoting each other."[96] The English demographer, Thomas Kirkwood of the University of Newcastle upon Tyne, concurs. "There is no reason why you shouldn't get greater defence against cancer and greater longevity."[97]

Raymond Kurzweil and the Singularity

Raymond Kurzweil is the Thomas Edison of our time. His many inventions draw on the field of artificial intelligence, including an optical-character-recognition program, a text for a speech voice synthesizer for the blind or visually impaired. He invented the first commercial speech-recognition system and with the singer Stevie Wonder, invented computer based musical instruments.

His efforts do not go unrecognized or unrewarded. He wins recognition and financial rewards, including the nation's largest award for invention and innovation, the $500,000 Lemelson-MIT prize, and the government's National Medal of Technology. He is a member of the U.S. Patent Office National Inventors Hall of Fame, and named Inventor of the Year by MIT and the Boston Museum of Science. Three United States presidents have honored him, and he has been awarded eleven honorary degrees.

All this may be but a preface to what may yet become Kurzweil's claim to fame or perhaps only to notoriety. He is now devoting his genius to the proposition that before long, people will have the capacity to live forever, or short of this—can enjoy much longer life spans. Ray Kurzweil takes 250 pills a day, mostly mega doses of vitamins, especially Vitamin

E. It is not that he is a pill-popper looking for a new high; he is instead, a serious scientist trying to live forever. "I think death is a tragedy, I think dying is a tragedy. And going beyond our limitations is what our species is all about."[98] To live forever, he recommends a diet of kale, seaweed, tofu, steamed broccoli, and bean sprouts, all washed down with green tea, which contains extra antioxidants to reduce the risks of heart disease and cancer.

Raymond Clyde Kurzweil was born in America to Jews who escaped Vienna as Hitler's Wehrmacht was closing in. His mother's father, a friend of Sigmund Freud, was a leading physician, and he got the family out of Austria. Ray's mother, Hannah, is a talented artist. His deceased father was a distinguished musician. The family initially lived in Jackson Heights, a lower middle-class section of New York, but later moved to a better neighborhood. Ray graduated from Martin Van Buren High School, one of New York City's better public high schools in its day. He later attended Massachusetts Institute of Technology. In 1965, while still in high school, he appeared on Steve Allen's *I've Got a Secret* television show during which he unveiled a piano rendition composed by a computer.

Ray Kurzweil was born into the upper levels of Europe's intelligentsia, and so his genetic roots may explain his intellectual genius. In matters of the body, however, Ray Kurzweil was not so lucky. His grandfather died of a heart attack. His father, at age fifty-eight, died of a heart attack. At age thirty-five, Kurzweil was diagnosed with diabetes. He was determined, however, to improve his odds of enjoying a long life.

Because insulin treatments did not work for him, Kurzweil drew on his skills as inventor and scientist to do his own research and determine how he could better help himself. In the process of trying to live longer, he came to believe that immortality might be a distinct possibility. Indeed, those who can stay healthy for the next twenty years, he asserts, may benefit from the scientific explosion he anticipates and so approach immortality.

By 2020, Kurzweil believes we will understand the nature of aging, cancer, and heart disease. He also believes life expectancy will continue to rise, and by 2030, life expectancy will be 120.

"Do we have the knowledge and the tools today to live forever?" In their book *Fantastic Voyage*, Ray Kurzweil and collaborator, Dr. Terry Grossman, admit that we do not have such tools at this time. However, because the rate of technical progress is doubling every decade, "the answer to our question is actually a definitive yes, the knowledge exists, if aggressively applied, for you to slow aging and disease processes to such a degree that you can be in good health and good spirits when the more radical life-extending and life-enhancing technologies become available over the next couple of decades ... we will have the means to stop and even reverse aging within the next two decades. In the meantime, we can slow each aging process to a crawl."[99]

Kurzweil and Grossman quote with approval the observation of Aubrey de Grey, a scientist in the department of genetics at Cambridge University that "Our life expectancy will be in the region of 5,000 years ... by the year 2100."[100]

Can our society cope with immortal life? With immortal life what shall we do with "the civilization that kills its youth on the highways and condemns its aged to pointless existence; that undermines marriage and the family; that either brings out the cancer that's in us or puts cancer into us; that profits financially from tobacco and sex and amphetamines and heroin and handguns and jet fighters; that dotes on violence; that admires predators, and that does not dare nor does it know how to renounce war. In short, our social arrangements militate against extended life. In this context ... it is just as difficult to change society as to change genes ... A biological breakthrough will force a new militancy, a new crusade, to 'Make the World Safe for Immortality.'"[101]

In his book, Ray Kurzweil boldly defines what he means. "The Singularity ... is a future period during which the pace of technological change will be so rapid, its impact so deep, that human life will be irreversibly transformed."[102]

Ho hum, so you say. So what? On deeper reflection, the Singularity has two major elements. The first is that the man/machine dichotomy has been breached. That is, Singularly incorporates nanotechnology, some no bigger than a single cell when combined with the biological nature of human beings, multiplies the human capacity for a super-long life of super creativity. The human and the machine, in cooperation with one another, make it possible for the rapid development of even complex ideas so that in an hour by the clock, humankind makes progress by the year. Moreover, such progress leapfrogs over previous developments so that immortality may be nearer than one thinks. If a machine can defeat Bobby Fischer at chess, what next is there for the rest of us?

As Yogi Berra asserts and as Kurzweil quotes with approval, "The future ain't what it used to be."[103]

Does God mean for us to die so that we may live? Kurzweil has no doubts. Life is good and death is bad. A long life is better than a short one, but immortality is best. Because immortality may be no piece of cake for the individual and society, shall we go along with Kurzweil or wrestle with our doubts?

The rapid progress of true science now makes, occasions my regretting sometimes that I was born so soon. It is impossible to imagine the height to which may be carried, in a thousand years, the power of man over matter. We may perhaps learn to deprive large masses of their gravity and give them absolute levity, for the sake of easy transport. Agriculture may diminish its labor and double its produce; all diseases may by sure means be prevented or cured, not excepting even that of old age, and our lives lengthened at pleasure even beyond the antediluvian standard.[104]
—Benjamin Franklin to Joseph Priestly (1780)

CHAPTER 6

The Golden Years

Death and taxes may not always be inevitable. Indeed, if the increasing number of mainstream scientists continue to ply their scientific craft so that what was once only in the imagination of science fiction, we may abolish or at least postpone the former well before we abolish or postpone the latter. That is, if a truly fortunate human being avoids being killed in accident or war, and avoids being stricken with a fatal disease, no one really knows what the maximum human life span, the number of years from death, a person might be. "Life span may be rising, but we cannot say what the limit is, or what may happen if medical advances, particularly genetic, succeeded in altering the figure."[105] Is there an upper limit to life? We do not know.

Despite all the data and despite all the techniques of demographers and actuaries, no one knows how long the luckiest of people might live. Science has suggested that based on the life span of other species, humans may expect to live six times longer than the years it takes to reach maturity. Thus, if it takes about twenty years for a human to reach maturity, the maximum life span for the luckiest of persons is about 120 years.[106] Since we have not yet reached that life span, clearly it is a marker along the way to eternal life. But whether many will reach such a life span in the current generation is certainly dubious.

> "This uncertainty about possible length of human life is of considerable symbolic significance for elderly people who live in a state of perpetual doubt. They never know how much time

they have, whether to begin this, or to promise that, whether to plan for several years hence, or at most for a month or two. They cannot tell if they should cling to their capital, persist in habits of saving 'just in case,' or should they spend while the opportunity lasts. They have great difficulty in deciding how much they should give away in their lifetimes, possessions, as well as money, should money be at their disposal. They can never be confident that a particular relative or associate near them in age will be present at all at any point in the future. In such directions, as these the lives of all the elderly go forward in a state of hesitation and incertitude, however active and satisfying those lives may be …"[107]

Peter Laslett identifies the specific sources of anxiety among the elderly:
1. Fear of death.
2. Fear of senility (Alzheimer's disease)
3. Fear of life-destroying diseases including cancer and heart disease.
4. Fear of blindness, deafness, lameness, incontinence.
5. Fear of mental decline and physical debility and for the dependence they bring.
6. Fear of loss of beauty, fertility, potency.
7. Fear of inability to recall names, events, people, and experiences.
8. Fear of losing the capacity for enjoyment.
9. Fear of loss of mobility.
10. Fear of loss of earning power.
11. Fear of loss of status.
12. Fear of loss of spouse, siblings, children, friends, relatives.
13. Fear of loss of home.
14. Fear of the future.[108]

To these anxieties, Laslett offers a further anxiety, namely, the fear of living too long. "Shall I go on living for too long? Shall I be forced against my will and in defiance of my choice to stay alive?"[109]

These are real and understandable fears of growing old, yet aging has its rewards as well. Once again drawing on Laslett, these may be enumerated as follows:

1. For many of the elderly, pensions become available and perversely,

the aged enjoy a level of financial security they may not have known before.
2. The dreaded mental and physical diseases associated with old age need not happen and for most elderly, these traumas may be avoided.
3. Most elderly do not end their lives in nursing homes, hospitals, mental facilities, or similarly unpleasant surroundings.
4. Plastic surgery and cosmetics may restore a measure of beauty and sexual activity. But since far more women than men live into old age, finding partners for women is often difficult. Remarriage or lasting relationships may grow.
5. The elderly, free from the necessity of earning a living, can give into pleasures of creative and intellectual life: reading, painting, writing, traveling, and studying with one's peers.
6. Some elderly find new careers and accordingly, continue earning financial rewards.
7. There is a future for the elderly to anticipate and enjoy.

Often, elderly grandparents are better off than their children are. Unexpected affluence can enrich the lives of their grandchildren by helping to pay educational expenses, or by helping with the costs of music, art, or athletic lessons, or by offering a trip to Europe or an experience sleeping away at camp.[110]

Laslett concludes, "We have to conduct our lives as far as possible not simply in remembrance of our former selves but in the presence of our future selves. The one thing which is certain in a world whose uncertainties we have dwelt upon is that we shall nearly all have a future to cope with, endure, if that is what we have to do, and enjoy if we possibly can."[111]

What if we find the fountain of youth and a breakthrough in longevity takes place in our time or at least in the time of our children, grandchildren, or great-grandchildren? Are we prepared for it? What changes in life's rhythms might we anticipate? If living forever is unlikely, suppose we manage so that within the short-term future the life span begins to approach the biblical one hundred and twenty years. Are we ready for it? Can we deal with it? If the coming of old age is always a shock, death is always premature.

Will a young person of seventy wait contentedly while his 120-year-old boss clings to his or her job and appears to thwart the ambitions of those preparing to follow him or her?

If there are decades more of life than the conventional threescore and ten, when does adolescence begin and end? Would childbearing be postponed indefinitely? Would a substantially lengthened life span pit the young against the old? How would we harness the energy and imagination of the young if the latter's future is indefinitely postponed?

Bret Stephens asks in an editorial in *The Wall Street Journal* from which some of the questions above are based, "Would not the bulk of human energies turn toward coarse and selfish attempts at self-preservation?"[112]

Immortal Life

Since the time of the earliest humans, immortality was a goal zealously sought but unlikely found. Yet, Philip Gordon of Yale University seems quite sure of himself when he declares, "In approximately the year 2050, we will have sufficient knowledge to be immortal."[113]

James Watson, the American molecular biologist and Nobel Laureate, compared the goals of the Human Genome Project (HUGO), which successfully mapped the human genetic makeup with America's successful landing of a man on the moon, declares, "We used to believe that our fate was in the stars. Now we know it is in our genes."[114]

"Dust thou art, and to dust thou shalt return," once a biblical certainty, is now a scientific doubt.

A recent piece in the science section of *The New York Times* carried this intriguing headline: "Your Body is Younger Than You Think." In the first paragraph, the author, Nicholas Wade, writes, "Whatever your age, your body is many years younger. In fact, even if you're middle-aged, most of you may be just 10 years old or less."

In a new method of estimating the age of human cells, its inventor, Jonas Frisén, a stem cell biologist at the Karolinska Institute in Stockholm, Sweden, believes the average age of all the cells in an adult's body may turn out to be as young as seven to ten years. "Although people may think of their body as a fairly permanent structure, most of it is in a state of constant flux as old cells are discarded and new ones generated in their place … The entire human skeleton is thought to be replaced every 10 years or so in adults, as twin construction crews of bone-dissolving and bone-rebuilding cells combine to remodel it."

While many cells renew themselves, some or all of the cells of the cerebral cortex do not, and so people behave according to their birth age and not the physical age of their cells.

But if most of the body is so perpetually youthful and vigorous, why do we age? The answer remains unknown, but among the theories is that although stem cells are the source of new cells they eventually grow feeble. Dr. Frisén is inclined to test this hypothesis. "He hopes to see if the rate of a tissue's regeneration slows as a person ages, which might point to the stem cells as being what one unwetted [sic] heel was to Achilles, the single impediment to immortality."[115]

Catch-22

In 2005, the nation embroiled in controversy over what became known as the Terri Schiavo case. Terri was a severely ill, disabled young woman. After collapsing in her apartment in 1990, she lost self-awareness and was essentially brain dead. Under the circumstances, her husband had doctors remove a feeding tube that had kept Terri alive. Was this an act of murder, as some Americans thought, or an act of compassion, as other Americans believed? In the emotional clash that ensued, the national debate centered on what, if anything, constituted death with dignity and was it more loving to put Terri out of her misery as her husband insisted, or more loving to continue to care for her? What elicited emotional attention to the case was that in America, as in many mature countries of the world, many families were faced with the problem of caring for a loved one who may never get well, but may linger on the cusp of death for an indeterminate amount of time.

To respond to a problem, which becomes increasingly evident in an aging society, there has been an expectation that people should think through their own preferences about how they wish to be treated during a prolonged infirmity during the last stages of their lives. As in the Terri Schiavo case, "What would Terry want?" if she could express herself, became something of a mantra.

On January 17, 2006, in the case known as Gonzales vs. Oregon, the United States Supreme Court upheld the State of Oregon Death with Dignity Act. The law adopted in 1994, made Oregon the first and only state in the nation to authorize Oregon physicians to assist terminally ill patients who no longer wished to live, to end their lives without pain.

In a six to three decision, the court avoided ruling on the philosophical implications of the measure, but asserted that former U.S. Attorney General John Ashcroft had acted without legal authority when he asserted that physicians who prescribed lethal doses of controlled substances—illegal drugs such as marijuana, opium, heroin—would lose their licenses to prescribe other medicines to ameliorate pain. The United States Supreme Court held, with Justice Anthony M. Kennedy writing for the majority, that inasmuch as the regulation of health care was a major responsibility of the states, Ashcroft had acted contrary to the "background principles of our federal system."

Although the case was decided on the narrowest of interpretations, the ruling made it possible for any state to enact assisted suicide legislation. Should they do so? Physicians are trained to heal. Hippocrates, the Greek father of medicine, admonished health care workers to do no harm. Should it now become part of the arsenal of sound medical practice to administer lethal medications? Or, was Attorney General John Ashcroft right when he asserted that assisted suicide is not a legitimate medical purpose.

In Joseph Heller's classic novel, *Catch-22*, the hero, Captain Yossarian, petitions his superior officer to be excused from further combat missions because he believes that after numerous successful flights into the war zone, his luck may run out. His superior officer, however, advises him of Catch-22, which in the military manual provides that a pilot may be excused from further battle if he can demonstrate that his mind is unstable. But, says the superior officer, if one requests to be excused from dangerous missions, it is clear evidence that there is nothing wrong with one's mind and that permission to be excused must be denied.

The Oregon law makes it legal for terminally ill patients to request their doctors to kill them painlessly, and the patient, not a surrogate, must make the request to at least two physicians. Secondly, the patient must have a clear understanding of what his condition is, what his alternatives are, and he must appear to be thinking rationally. But it is difficult to contemplate that a person so ill can really think rationally. Much of the time, those in extremis are at least depressed. Moreover, no one rationally asks to be put to death. No one really wants to die. Thus, Catch-22, to ask to be put to death is an irrational request and demonstrates that the person making such an absurd request is not of sound mind and so has at least limited mental capacity to make such a request. Therefore, if the law is followed, the request must be denied.

Moreover, modern medicine provides alternatives for those who feel themselves near death. Numerous painkillers can make the end of life tolerable. Some physicians, however, have been accused of making limited use of such medicines for fear of making the patient addicted to the controlled substances. But, of what importance is addiction if the patient is at death's door?

In addition, there is hospice or palliative care, which eases the burden of both patient and caregiver. In palliative care, the physician no longer seeks to make the patient well. Instead, doctors seek to make the last six months or so of a patient's life more comfortable and as painless as possible. Hospice care, moreover, offers much needed support for a patient's family as they prepare themselves for the death of those they love. "Thanks to medicine's prowess in sustaining life on the edge, it is harder than ever to know when it is 'time to die.'"[116]

A Perfect Social Storm

The medical ethicists Eric Cohen and Leon Kass recognize that "Americans worry about the soaring costs of Social Security and Medicare, the collapse of private pensions, the shortage of good nursing homes, and the potential clash between the young and the old over resources and priorities." But, they continue, "Our deepest worries are personal: we dread spending our final years in a degraded state, resented by caregivers or abandoned by loved ones, of little use to ourselves, never mind to others."[117]

According to a Rand study, 40 percent of deaths in the United States are now preceded by a period of enfeeblement including debility, dementia, and sometimes in such a protracted state of being unable to care for themselves for as long as a decade.

More than four million Americans have Alzheimer's disease, which deprives them of self-control and self-awareness. By 2050, the number of Alzheimer's patients is expected to rise to over 13 million.[118] Alzheimer's disease plays no favorites. Rich or poor, physically fit or frail, vital and active, Alzheimer's is the Grim Reaper of our time. Former President

Ronald Reagan suffered from Alzheimer's for an extended period before he died, and actor Charlton Heston was a victim.

How shall we deal with the issues derived from the dilemmas of such a scenario? How shall we maintain men and women who live longer but retire from work earlier?

Who should be given preference on the job, the sixty-five-year-old with the vigor of a thirty-five-year-old and wants to work, or a younger person with the same chronological age?

How shall we enable a family to provide care when the size of families is shrinking, and many live far from the homes of the frail elderly?

How shall we relieve the shortage of good nursing homes required more than ever in the face of an aging and frail population?

How shall we reward those caregivers who feed and toilet those unable to cope with their own personal hygiene?

If in our search for immortality, centenarians become the rule rather than the exception, how will we cope with the additional years of life?

How will people use their extra years of life? Will they work more or play shuffleboard?

Will children be willing to wait indefinitely for their legacies?

Will colleges and universities, museums and hospitals that depend on bequests from the wealthy likewise wait indefinitely?

If not all can benefit at once, who will benefit first from the gift of additional years of vigorous life? If families living on New York Park Avenue are among the first to benefit, will the people of disease-tormented continents be willing to wait happily and patiently for their turn?

The gift of life may be likened to the *Titanic,* a huge transatlantic vessel supposed to be the fastest then afloat, only to crash into an iceberg and sink in one of the greatest sea tragedies of all time. With lifeboat space limited, it was supposed to be women and children first into the lifeboats, but such was not to be. The issue of whose life was worth more was raised in a raw clash not only between men and women, children and adults, but also between rich and the poor. No group aboard that ship acted with complete virtue. Were there to be a rapid expansion in the direction of eternity, one would need to do more than to rearrange the deck chairs.

In Chinese calligraphy, the symbol for crisis consists of two figures, one of which standing alone means danger, the other by itself means opportunity. Prolongevity, eternal life, life extension, whatever we call it, is a mixed blessing. We will have to know how to reorder our lives to minimize danger and welcome the opportunities afforded us by the gift of eternity.

Indeed, it would become necessary to develop a new image if eternity. How should we do so? As Cohen and Kass assert, "All this creates a perfect social storm."[119]

There's a far land I'm told, where cigarette trees and lemonade springs abound; the hens lay soft-boiled eggs; the trees are full of fruit, and hay overflows the barns. In this fair and bright country, there's a lake of stew and of whisky too. 'You can paddle around 'em in a big canoe.' There ain't no short-handled shovels, no axes, saws, or picks. It's a place to stay, where you sleep all day, 'where they hung the jerk that invented work.

—Sebastian de Grazia [120]

Chapter 7

Making the World Safe for Immortality

In *Deuteronomy* may be found the biblical injunction, "I have set before thee life and death, therefore choose life that you may live." In secular terms, that choice is between meliorism and apologism. According to Gerald Gruman, a physician as well as an historian, meliorists are those who take the view that we can do much to improve and prolong life, and even thwart death. Apologists believe there is a moral imperative to die. During World War I, an American sergeant exhorted his men to charge during the violent battle of Belleau Wood. "Come on," the sergeant urged. "Do you want to live forever?"[121] Such is the nature of the human condition that hope springs eternal and no one wants to die.

Look Ma! I Still Drive

A rite of passage to adulthood for the sixteen-year-old is a driver's permit, followed by a state administered test of driving performance, and then the award of a license to drive. For the young person, a license to drive is more than evidence of driving competence, it is also a measure of growing esteem among peers, and an expression of personal independence (even if they must, at least for a time, ask their parents if they can use the family car). Teenage drivers, judging from the number of automobile accidents, are among the worst drivers in the nation. In second place, however, is the dismal record among elderly drivers.

In an Op-Ed piece for *The New York Times*, an orthopedic surgeon and triathlete seriously injured by the erratic driving of a seventy-five-year-old man, cites the following:

Motor-vehicle injuries are the leading cause of injury-related deaths among 65- 74-year-olds and are the second leading cause after falls, among 75- to 84-year-olds. Older drivers have a higher fatality rate per mile driven than any age group except drivers under 25. The American Medical Association estimates that as the population ages, drivers aged 65 or older will eventually account for 25 percent of all fatal crashes.[122]

The inexperience of new drivers, people driving under the influence of alcohol, and speeding, are the leading causes of automobile accidents. Among the elderly, however, neither alcohol nor speeding account for most accidents. It is their impaired vision, slow response times at critical moments, reduced peripheral vision, and osteoarthritis, which means they may not be able to turn their heads or bend their knees to react appropriately in traffic, that cause accidents. Elderly drivers are rarely required to take another performance test during their lifetimes, and not all states require a vision examination when an elderly driver's license comes up for renewal. Should elderly drivers, say those over sixty-five, be periodically required to take a driving performance test, and if they fail, should their licenses be denied them? The orthopedic surgeon urges that elderly drivers be retested every five to ten years, and if their performance falls short, their licenses should be withdrawn. So it would seem, but wait a minute.

Just as the award of a driver's license is symbolically a token of emerging independence for maturing teenagers, taking a license away from elderly drivers denies the elderly driver the independence to which he or she has become accustomed. If life is independence and death is life's end, then, as viewed by the senior citizen, dependence is purgatory.

For a majority of the elderly, driving offers the independence that growing old seems to take away. Is a return to dependency a social good? Given that, many and probably most elderly drivers live in areas where public transportation is limited or unavailable, who drives the elderly to the movies, to concerts, to museums and libraries? Who drives the elderly to restaurants or to shop? Who drives the elderly to the doctor? To take their licenses away seems a prudent act in the interest of safety on our streets and highways, but the price paid is to establish an underclass of women and men who can no longer drive.

Moreover, in states like Florida and Arizona, where many elderly drivers form a substantial voting bloc, requiring periodic retraining and retesting of senior drivers is the third rail of state politics. How we deal with drivers licenses for the elderly becomes a metaphor for how we or our children or grandchildren will live in a protracted life span. As Joe Coughlin, who directs the age lab at the Massachusetts Institute of Technology correctly observes:

"Transportation is a lot more than simply going to the store a n d going to the doctor's. This is the way that we maintain the connections with all those little activities that when you put them together we call life. And that transportation, that driving, is the glue that holds our life together." [123]

Joe Coughlin recognizes:

"We are an aging country. And all of a sudden all those things that we take for granted every day, like driving or cleaning our home, going out shopping the natural aging process does begin to make those things more difficult. As a nation, we are not prepared for the very success, the greatest success of the last century- that is, longer life. And so the transportation issue points to larger questions. How will we work? How will we play? How will we get around? How will we live tomorrow now that we live longer?"[124]

The Demographics of Aging—the Twenty-first Century

"Few Americans realize that their country is in the midst of a demographic revolution that, sooner or later, will affect every individual and every institution in the society. This revolution is the inexorable aging of our population."[125]

In this section, we will venture to report estimates of the growth of the elderly among us in the United States. For statistical purposes, those sixty-five or older are considered elderly irrespective of their health or fitness.

*In 1920, every twenty-second American (4.6 percent) was sixty-five or older. By 2020, at least every fifth American (20 percent) will be elderly.

*The baby boom generation (born 1946–1964) will swell the size of the eighty-five and over population.

*Between 2010 and 2030, the proportion of elderly will increase sharply unless large-scale immigration will lower the proportion of elderly.

*In 2050, 55 percent of the 65-plus population will be over 75. By the same year, problems of providing health and social services, Social Security, adequate housing, and satisfying jobs for the elderly will be great and growing.[126]

*By 2050, the median age of the population will be forty-nine. About 29 percent of the population will be over 65, and only 18 percent will be below 20. Death will be much more frequent than births, and the number of chronically ill persons will vastly increase. However, by 2050 life expectancy may reach 100 years.[127]

Of Time, Work, and Leisure

In 2006, young French adults demonstrated against a government proposal, which seemed to provide youth with greater opportunities in their initial jobs at the expense of job security. As one of the banners of the young protesters asserted, "We want to work less so that we can live more." Living more and working less will become the mantra of many people in the industrial countries of the world and expand to less developed countries with more or less intensity. To use the extra years of life that science may make possible in indolence is to doom a people and a nation to decrepitude and decline. Is indolence, however, the only alternative to work? If people must work if they are to be human, but there is little work to do, what then for humankind?

With a longer life expectancy and a strong preference for retirement rather than work, how shall people use the time in which they do no work? They will use their time in leisure. If leisure is defined as neither sloth nor indolence, leisure must be reinvented, and we have an excellent opportunity to do so. At least such is the view of Sebastian de Grazia, chief author of the study, "The Twentieth Century Fund, 1962," *Of Time, Work, and Leisure.*

A person at leisure is not one who is idle. Leisure, de Grazia insists, is neither play nor indolence. If leisure is defined as "freedom from the necessity of labor," and if Americans prefer retirement to work, then the challenge to Americans is to convert free time into leisure, not work. "Wisdom is the virtue that cannot appear except in leisure."[128] Since the elderly have finished their work, but still have many years of vital living to do, what shall they do with their remaining years?

During work years, there are goals to achieve, careers to navigate, careers to be made, money to be earned and invested, saved or spent.

Thus, are the workers of the world fulfilled? But how shall the leisured be fulfilled? Since the elderly have no future at work, they can use their leisure time to become vital contributors to the future of others. Thus, "the elderly of any society can be said to be trustees for the future."[129]

Those well placed in government or industries have traditionally occupied the leisure class.

An expanded life span offers an opportunity to democratize leisure, that is, to make a life of leisure available to those to whom leisure has been denied. A leisure society elevates and offers new opportunities—for those who may not have had it before—the time to read, write, or study, to learn to paint or sculpt, develop craft skills that might otherwise be lost, to listen to concerts or to learn a musical instrument, or master a foreign language especially, perhaps, a more esoteric one. They may assist with non-governmental organizations in offering help to the world's poor, to those sick with AIDS or malaria or malnutrition. They may offer their services to libraries or museums. Or, they may return to school and study topics they had perhaps once scoffed at, but now appear to have more relevance. Thus, *not* working may be the good life.

To provide such opportunities requires a public sector that encourages a leisure ethic so that the school, the library, the museum, not the bank, the farm, or the factories become the focus of a leisure life. Leisure, and the means to experience it, has ceased to be the monopoly of the elite. Work may enrich, work may even ennoble, but leisure perfects and "In this lies its future."[130]

If leisure, not work, is central to a prolonged life, then a strategic role for education (schools, libraries, museums), needs to be found. Schools can no longer be confined solely to the preparation of the next step in the work cycle. While schools will continue to function to provide the young with essential skills needed to be a literate person, or to prepare for professional careers or vocations, schools must offer a greater number of choices than currently are available. Schools must no longer be the monopoly of the young. Nor can schools monopolize education, but must offer diverse age groups the support they require while sharing an obligation to work with other agencies offering educational enrichment. The central question for a leisure society is, "How can the dramatic rise in life expectancy become the basis for new social productivity—for a genuine abundance of life?"[131]

With the rapid growth of the elderly population, will the young now support the old in making it possible for the elderly to enjoy life's abundance as they provided for the young? The enlargement of the elderly among us offers a unique opportunity and challenge to organize our institutions to make a continuing and cooperative relationship between the young and the old. Indeed, crabbed age and youth must learn to live, work, and play together.

Am I My Brother's Keeper? What One Generation Owes Another

"In 1920, a ten year old in the U.S. only had a 40 percent chance of having two of his or her possible four grandparents alive. Today, that figure is 80 percent. In the 1950s, the average age of admission to a nursing home was 65. Now it is closer to 81."[132]

Today, growing old is growing better. Nuclear families of the past are more myth than reality. The era of multigenerational families is here. Many older couples live long lives and have the opportunity to see grandchildren and even great-grandchildren grow and prosper as self-directing adults. This is a blessing not to treat lightly. In the United States, unpaid caregivers— members of the family such as spouses and children—are relied upon to help when the elderly become infirm. Will family members continue to do so over long periods? With fewer children to rely on in old age, what is the responsibility of the young for the old?

In most of human history, the old felt a social obligation to the young. They provided food, clothing, shelter, and protection from whatever hostile forces, human or divine, that befell them. Mostly, the elderly taxed themselves heavily to provide elementary and later secondary schooling for a nation's youth. When Horace Mann in the 1830s devised a model for the common school, he had to convince those who thought otherwise, that it was prudent to tax a family who had no children so that the children of large families could attend school. "Why tax me," so the argument went, "to pay for the schooling of children of someone else's family?" As we know now, the issue was resolved, and here in the United States, the most egalitarian public school system was established.

Today, we face a different circumstance. For many centuries, the shift in resources from the old to the young was debated against an agreed upon assumption that the old had an obligation toward the youth. Today, the issue is whether or not the young has an equal obligation toward the old, and if so, how best to fulfill that obligation. It is no less a transformation in thinking comparable to that of the Renaissance, which made available a treasure trove of scientific, musical, and artistic effort. It is a transformation of faith, comparable to the Reformation, which provided opportunities for religious pluralism, as the Industrial Revolution made work less burdensome while raising living standards. Just as these social transformations required centuries, so too will a shift in thinking be required to make a longer life, if not immortality, a rewarding one for the ever increasing numbers who are adding years to life so that they might also enjoy life in their later years. One may identify the paradigm shift between the young and the old, as the Fourth Social Transformation. This shift in perspective is a debate in which we remain engaged and which will not be readily resolved. We ignore it at our peril.

Close to the heart of the twenty-first century debate over comprehensive health care reform, is precisely the question of what do the old require of the young, and perhaps even more importantly, what are the obligations of the young to the old. Many of those who oppose comprehensive health care reform do so because such massive reform will "impoverish our children," or will saddle them with a burden of debt from which they will not recover. Yet, did not the now elderly cheerfully burden themselves to nourish the physical, social, and career needs of the youthful of their day?

Comprehensive health care reform has become an icon for the state's obligation to their citizens just as it was a century or more ago. The debate at that time was whether education at public expense was a responsibility of the government.

Precise definitions of who is old and who is young are not especially helpful, because the young, if lucky, will one day become old.

In the Fourth Social Transformation, the issue is not simply how best to allocate scarce resources between the young and the old, but to recognize that the old and the young are part of a single continuum. As an extended life span becomes a reality for a greater number of old people (say, over sixty-five), the shape of that life span needs reexamined. Aggravating matters is that in the United States, as in most developed countries, the pool of unpaid caregivers is declining because of smaller families.

A major issue in the Fourth Social Transformation is how to make the transition from unpaid caregivers to paid caregivers. In Japan, the state provides for the care of the aged in institutions or at home as needs and circumstances dictate. In the United States, the direction is relying on for-profit managed care institutions. In the United States, there are numerous choices for the elderly including home care, and varying degrees of assisted living depending on age and the nature of the infirmity. But, will the private sector provide the level of support the elderly may require, and if the public sector is called upon, will the young tax themselves more heavily to support the elderly?

A Faustian Bargain

At one time, children were widely viewed as miniature adults—they were thought of that way and treated that way. It was not until Jean-Jacques Rousseau wrote that childhood was a special time of life and children had special needs by way of shelter, food, clothing, and education. They needed to be brought up carefully if they were to grow into respectable maturity. In a parallel way, women and men at say, eighty-five, are not merely older versions of what they were at say, age fifty-five. Anthropologist Margaret Meade said it well when she declared no one would live his or her life in the kind of world in which he or she was born, and no one will work in the kind of job he or she performed in his or her maturity. An octogenarian is a different person from a middle-aged person and so has different needs. The socio-political-economic model appropriate for the former may be a disaster for the latter.

One cannot expect that a young person starting a new job at age 25 will know what his or her needs will be at age 65, 85, 95, 105, or in the biblical 120. The shape of the life span requires ever-greater opportunities to choose school or alternative work, alternative forms of housing, and living arrangements. The marriage vows taken "til death do us part," become very real as one of the newlyweds may outlive the other by a substantial period. Moreover, the very institution of marriage may need to be refashioned in that the happily married couple at age 25 may no longer be happy with one another at age 65. They may not want to remain married over an extended life span of 85 or 95, to say nothing of 120, which some life-extension researchers say may be achieved during the next half-century or so. Multiple marriages may become the rule, rather than the exception. If people live longer, they may need to work longer to support themselves over an extended life span. But, who hires the elderly? Do the elderly stand in the way of career opportunities for the young?

Since members of the United States Supreme Court serve for life, is it desirable to have court justices serving for a hundred years?

In an aging society, health care costs may increase, But must they? Perhaps as people live longer, the onset of diseases such as diabetes, arthritis, Alzheimer's, Parkinson's, and heart and prostate problems may be delayed and some people may live long enough to benefit from a cure. Contrary to conventional thinking, as life span expands, the elderly may become healthier and health care costs may go down, not up!

In the competition for initially scarce cures of chronic diseases, is it fair for the rich to have the first chance to obtain the cures? Should medical advances become, at least for a time, initially available to affluent nations?

Dr. John Harris, a bioethicist at the University of Manchester, England, is not troubled by these questions. He believes that scientists have a moral duty to extend the human life span as far as it will go, even if it means creating beings that live forever. "When you save a life," he declares, "you are simply postponing death to another point. Thus, we are committed to extending life indefinitely ..."[133]

Adam and Eve, expelled from the Garden of Eden, lost their claim on immortality. But the search for immortality still goes on. Should it?

Dr. Leon Kass believes that the pursuit of immortality requires a Faustian bargain in that a person who lives to 500 years is no longer human. A person who will live that long will lose the meaning of life and will fail to cherish each day that a mortal life span requires.

Daniel Callahan, the distinguished medical ethicist, shares many of Leon Kass's views and believes that even if we live to be 500, "we will still be human beings."[134] Nevertheless, in Callahan's judgment, the search for immortality should be approached gingerly and kept at arm's length since we do not know the kind of society we want, nor do we really know the kind of society we will get. Callahan insists, "We had better not go anywhere near it, until we have figured those problems out."[135]

Brave New World

If drugs can provide humankind with happiness everlasting, is that as bad as Aldous Huxley observes in his 1932 dystopian novel, *Brave New World*. Drawing on the lines from Shakespeare's *The Tempest*:

"O wonder!
How many goodly creatures are there here!
How beauteous mankind is!
O brave new world,
That has such people in't!"
(Act V, Scene 1)

In his book, Huxley portrays a world society seven hundred years into the future from the time he wrote the work. A time in which every need is met, every sorrow is seemingly overcome, there are no further goals to seek, and no social mandate other than the obligation to be happy. War and disease have been overcome, as have the seven deadly sins, which are neither deadly (since disease has been overcome) nor sinful. There is no sin of envy, greed, gluttony, arrogance, or lust. Indeed, sexual license is the rule rather than the exception, because to conceive a child in the womb is considered rather gauche, to say the least. All children are products of in vitro fertilization and are subject to genetic engineering. Soma is the preferred, indeed required, antidepressant. Aging has been abolished as youthful sixty-year-olds willingly take an overdose of Soma and die without complaint, without anxiety, without remorse, and without a prolonged and painful terminal illness.

The world Huxley depicts is one of sloth, which addiction to Soma makes humankind able to deal only with the superficial and have interest in the celebrity who is famous only because he or she is popular. Under the influence of Soma, friendships are tentative, family relationships loosely knit, and entertainment superficial. Little is newsworthy, most news is not to be taken seriously since everything, and anything can be overcome. What is important is only the fun you have in and with a healthy body.

Huxley thinks it is terrible. Is it? Need it be?

President Dwight Eisenhower cautioned against an emerging military-industrial complex. In *Brave New World*, Huxley portrays a world where the danger that the brave new world may be dragooned by a biomedical industrial complex. Ask any senior citizen, most are beholden to a score of pills they must take daily to live. But the pills that alter the psyche, like Huxley's fictitious Soma, are those that are most dangerous, because they alter mental perceptions. Yet, are pills that build self-esteem any more to be shunned than diet pills, which may make obesity a disease of the past?

Huxley's error was to pour the new wine of a world order enhanced by science and medical technology into the old bottle of society as it existed in 1932 when the book was first published. Those were indeed dark days for America, and the world mired in the depths of the economic depression. So bleak were the prospects for the economy that new institutions had to be developed to cope with the emerging problems. Thus, we came to Social Security benefits, unemployment compensation, farm subsidies, bankruptcy protection, and bank regulation, safety regulation in mine and factories, minimum wage laws, and a more effective system of social justice. *Brave New World* is a reflection of Huxley's pessimism of where the world will be seven hundred years into the future. By taking such a long time frame, his views make for good reading and stimulating discussion but remain out of the realm of thoughtful analysis. His views can be neither proven nor disproven, and at the least, we have the privilege of either accepting or rejecting his views. However, his views are interpreted; he offers only a science fiction blueprint of a gloomy world brought about by our own successes.

If, in the rapid emergence of biotechnology and with it the opportunity to enhance life's quality even as it is lengthened and achieves serenity rather than anxiety, shall we abandon the effort. Is it beyond the mind of men and women to order society to derive the benefits of science without destroying ourselves in the process? Is it humankind's destiny to be unhappy? Is real life better than bliss eternal? "

Huxley neither knows nor understands the glory of what lies ahead. A utopian society in which we are sublimely happy will be far better, not worse, than we can presently imagine. And it is we "who neither know nor understand the lives of the god-like super-beings we are destined to become."

*And now to all us women may Christ send
Submissive husbands, full of youth in bed,
And grace to outlive all the men we wed.*
—Geoffrey Chaucer (1340–1400) "Wife of Bath's Tale" in *Canterbury Tales*

Epilogue

Alternatives to Death and Dying

In 1967, the Beatles released their popular ballad, "When I'm Sixty-Four." In 1967, life expectancy among American men was sixty-seven; women lived about seven years longer. The projections of the Beatles about life at what they believed were the end of life possibilities, pales by comparison with the startling projections demographers make today. We live in a time when distinguished demographers are suggesting a prolonged life span; something beyond eighty-five appears a probability in the next generation or two. Yet, we also live in a time when wanton death through war, terror, and plague is all too much with us. Death from the ravages of time appears remote or at least postponable. Despite the tongue-in-cheek title of this book, no one will live forever, yet death is always premature, and like the oncoming of age itself, death is always a shock and "always remains a misfortune."[136]

Just as one is shocked to find upon looking through the bathroom mirror that we are old, one is equally shocked to realize that death will come before we have finished our life's work.

"Sigmund Freud saw the idea of immortality as supreme among civilization's hopes."[137] Yet, in the resurrection of Jesus, or in the promise of life ever after in the rapture anticipated by fundamentalist Christians, religion holds out a promise of eternal life. Whether it is the soul that never dies, or reincarnation, or salvation, or achieving nirvana, or the transmigration of souls, religion seems to give us comfort that some part of us will not die.

In our time, suicide terrorists, whatever their political motives, seek also immortality through death. The suicide terrorist sees in death a connection with immortality not unlike the hara-kiri or ritual suicide among Japanese samurai. Or the Japanese kamikaze pilots who crashed into enemy aircraft to end their lives and the lives of the enemy, or the suicide of victims of Nazi concentration camps who sought immortality in death at their own hand and not that of the enemy.

On the corporeal level, we can learn to starve ourselves to live longer. As reported in *The New York Times* on September 22, 2000, "Biologists report today that they have shown precisely why a calorically restricted diet prolongs lifespan at least in a lower organism. Laboratory rats and mice live up to 40 percent longer than usual when fed a diet that has at least 30 percent fewer calories ..."[138]

The time is not yet ripe for one to starve one's self in exchange for immortality, but apparently, a low caloric intake may contribute to life's longevity. Maybe by eating less you will not live longer, but it may seem that way!

After nine years of disappointing research, Dr. Leonard Guarente, professor of biology at the Massachusetts Institute of Technology, believes he can explain the effect of how a low caloric diet prolongs the life of yeast cells. What may be true in yeast cells may be true in human cells. However, he does not jump to this conclusion just yet. If a drug could be found to work on humans the way a 30 percent drop in normal calories works on rats, "people would start enjoying a maximum life span of 170 years, most of it in perfect health."[139]

Michael Rose, a University of California-Irvine evolutionary biologist in his research on prolonging the longevity of fruit flies, sees a bridge to his new study of immortality. "It is an Einsteinian revolution compared with what we used to do."[140] While immortality may be generations away, its achievement is a distinct possibility. At least, so he believes. "A scientific land rush," he asserts, is underway. It focuses on three broad areas: the genetics of aging, techniques to immortalize cells and tissues, and exploitation of the basic modeling clay of our bodies-pluripotent stem cells."[141]

In a graph entitled, Immortality Reality Check, science journalist Brian Alexander notes:

"If you are 100: You have a fighting chance of hitting 110.

"If you are 70: You have a respectable shot at 100. Your chances are better if you are a woman.

"If you are 40: Most of you will push 85. More than 1 million of you will live to be 100 or older ... a significant number of you will live past the current maximum human life span, possibly to 135.

"If you are 30: Advanced plastic-implant and tissue engineering will give you access to augmentations and biologically match replacement parts. Some of you will even live to see 2100.

"If you are 10: Immortalizing therapies will be available by pill and injection by the time you hit 40.

"If you are minus 20: Bingo! By the time you're a gleam in your parents' eyes, they will be able to choose germ line engineering tinkering with sperm and egg to select for extended life span. Many of you will be the first substantially augmented human beings. Can-do researchers figure that all these bonuses will combine to make life spans of 500 or 1,000 possible. But just to be safe, write out that last will and testament no later than your 200th birthday."[142]

Appendix A

One Hundred Plus Ways to Age Gracefully

1. Look forward to growing older

Many people resist growing older. While aging may not be welcome and needs to be resisted, one must recognize that age has its prerogatives, and inasmuch as age is better than the alternative, it must be accepted, and in a sense, welcomed. We must recognize that there is no fountain of youth (not even with the best of plastic surgeons).

2. Be careful of what you remember

Memories tend toward exaggeration. That is, the triumphs are recalled as overcoming inordinate odds, while disasters are likewise remembered as hitting bottom. Neither is true. If we remember disasters that have overtaken us—careers not altogether fulfilled, roads not taken, mistakes made, friendly overtures turned aside—we are courting depression. Forgetting is not all bad.

3. Live in the present

St. Augustine said, "Time is a three-fold present. The present as we experience it, the past as present memory, the future as present expectation." Those of us who were born before the computer age will never become a genuine part of it, but we need to master it, along with the ubiquitous cell phone. Things get away from us. Jogging shoes, backpacks, water bottles, cell phones, e-mail, are all part of the present and make up no part of my past. Yet, there is no denying that the present is all we have and we need to cope. Moreover, it is not true that an old dog cannot learn new tricks. They can and we can too!

4. Beware, or be wary of children

Children are in the way. They are in the way when an older man or woman seeks romance and companionship. They are in the way when one needs to make out a will and may wish to provide for other than children. Children often cannot abide an older adult acting young. To children, older adults ought to get out of the way and die quickly. Thanks, but no thanks.

5. Don't sweat the small stuff

Sometimes I get up in the morning and nothing hurts me. But I don't worry about it, because I am sure it will pass. That is, as we grow older aches and pains develop, joints stiffen, the walk slows, and we are no longer in a hurry. It takes us all day to write a letter. Remember when twenty or more letters a day was the norm? Should we choose to run to a doctor every time there is an ache, we can, but beware that the search for a cure may be an addiction and the doctor a false messiah.

6. Nothing to fear but fear

Winters in New York seem colder, snowier than usual, and we are fearful of going outside. Mostly, we fear falling and breaking a hip or a limb. We need to take precautions. However, the very fear, the timidity, the loss of confidence, contributes to what we fear the most. I used to ride a bike and drive a car. Today, I fear both biking and driving. It is simply not worth an accident. When I was a kid, heaven was having a car. When I finally got a vehicle, heaven was finding a free parking space. Later, heaven was having an expensive garage in Manhattan. Today, heaven is getting a taxi or better, a limousine with a driver. Take precautions and realize there are limits. *Fear of tomorrow can cause you to miss moments of happiness that can be shared today.*[143]

7. Keep moving—Exercise

Despite potential fears, it is of enormous benefit to continue to exercise and not give up on the activities that you began in your youth. For example, I used to be a modest jogger—no more than three miles each morning, irrespective of the weather. Today, I do not jog any longer because I tore a ligament in my right knee. I miss jogging, but I keep moving. I go to an athletic club twice a week where a personal trainer works with me. I use a stationary bike consistently.

8. Prepare to live alone

When my wife died, it was like losing my right arm. I could not file tax returns, keep household records, or get what was due me from medical insurance. I could not balance my checkbook or boil an egg. (At best, my soft-boiled eggs are medium hard.) In any case, I acquired survival skills. I can broil a steak or salmon, open a can of soup, and shop at a supermarket with some degree of confidence. By the way, the supermarket is a good way of meeting and finding companionship of the opposite sex. Nothing came of chances I had, but I am glad I had the chances! (One woman wanted to come to my apartment and show me how to broil salmon.)

9. Be ready for anything, because anything will happen

I was once a member of a group called, The Future Society. The group never made a claim that it could predict the future, but we believed we could figure out some alternative futures and sound a clarion call to the nation to do likewise. I do not know what happened to the society, but we felt ready for anything, because the unanticipated is what is likely to happen.

10. Old age is not for sissies

It takes courage to grow old, and for most of us, old is not attractive. What with wrinkles, fat, and scars, we are often reluctant to see our reflections in a mirror. Infirmities abound, but more than that, we begin to examine at a level that squares with our education and capacity for introspection, the nature of our lives. We ask ourselves, "Did we live the good life?" Indeed. "What is a 'good life,' and how do we know we have had it?" At this point, I don't know how I would answer this question, yet alone generalize for others.

Since old age is not for sissies, what does one do for an encore? *There are miles to go before I sleep/And promises to keep/And promises to keep.*

—Robert Frost

11. Grief is good

For many years now, I have looked at the obituary columns of *The New York Times*. In nearly every issue, I see notices of men and woman my age

that have died. I have also lost loved ones and dear ones and near ones, and it is necessary and good to grieve over them. A woman of years in my group of friends lost her second husband whom she loved dearly. Allegedly, following her psychiatrist's advice, she decided she was going to make a life for herself. The evening after the funeral, she went to a sumptuous restaurant with some friends. While excessive grief is a sin, resisting grief is a sure road to depression. *Rage, rage against the dying of the light and not go gentle into that good night.*
Dylan Thommas

12. Curious George or Tom or Dick or Mary

It is important to remain curious, to try new things. This does not mean that you need to make a foray into the bold unknown, but in modest ways, like read a new or an old book of a genre you might not have thought of before. Visit a new museum or see a new exhibit, a new play, a concert, or a movie. Learn a new game, bridge perhaps, or backgammon; take music lessons. Get out of the rut.

A new restaurant, even if it turns out to be a disaster, may be good for you. Travel to a part of the world you might have otherwise ignored. In this age of terror, airline travel can be viewed as an obstacle course going through checkpoints, luggage inspection, and electronic gates. Nevertheless, this too is an experience. Obviously, prudence is called for and anticipating difficulties is essential. But one cannot allow oneself to be utterly thwarted.

A tribute:

On his birthday, former President George H.W. Bush boards an airplane and takes an annual parachute jump. I am not suggesting that readers take parachute jumps, but even in old age, one can try to recapture the skills one developed in one's youth.

13. Medical tests and other temptations

As we grow older, we increasingly worry about our health, as well as we should. But the dilemma of the twenty-first century is that the tests required to find out what, if anything, is wrong with our health, are complex, expensive, not without risk, sometimes painful, and often inconclusive. Dr. H. Gilbert Welch wrote a book entitled, *Should I be Tested for Cancer?* This professor of medicine at Dartmouth Medical School suggests by way of his subtitle: "Maybe Not and Here's Why." Tests may be wrong, yield false positives, and create needless worry. There are times, of course, when tests are necessary and indeed mandatory, but you need to approach them with healthy skepticism. Testing for cancer may save lives, but not always.

14. What friends are for

As one ages, friends tend to slip away. That is, some die and others grow infirm, but to withdraw from a social life is unacceptable. Friends share common interests, relieve loneliness, and offer companionship. For some, it is not easy to find new friends to replace those who are no longer available. One of the more useful ways to make new acquaintances is to return to school and participate in the vast number of study groups and other activities that colleges, universities, libraries, museums, and religious organizations provide. Some of these may be costly, but others are free.

15. Adopt a pet

This is not for everyone inasmuch as pets require care. But to the extent that a pet can be cared for, a pet often provides amusement, companionship, and a measure of contentment. Dogs require substantial care, cats perhaps less so. Birds need attending to, and tropical fish provide serenity in an ever-changing fish tank.

16. Do one naughty thing each day

My naughty choices, in order of preference are ice cream, a chocolate bar, a hot dog, a scotch or martini, or a glass of red wine with dinner.

17. Exercise the brain

Just as it is vital for a gracious elderly person to be physically active for as long as one can, it is important to get involved in some intellectual challenge. Read regularly, stay current on controversial, political, and social issues, and join a book or issues discussion group are ways of exercising the brain. Playing bridge or backgammon at a decent level of competence, and doing crossword puzzles or other word games may likewise keep one's brain active. If your voice is better than your mind, join a choir and sing! Travel can also, be a source of brain power development, especially if you join the ever-increasing number of groups who traveling abroad to destinations sponsored by schools, and colleges, and museums. Some of the journeys have a bit of an edge to them, but therein lieslies the danger, the fun, and the learning.

My personal adventure at age eighty-two:

In 2005, I joined a group on a small cruise ship, which included a stop for several days to visit the ancient Roman remains at Leptis Magna in Libya. Our ship was the second one to visit this once forbidden port. We had a group visa, so we had to stick together in this country, which the United States State Department lists as a terrorist state. My understanding is that for reasons unknown to me, Libya has suspended these tours. The experience was a thrilling one for me because we were among the first group to visit a formerly forbidden land, and because of the real or imagined danger of straying or doing anything of an independent nature. One needs to choose one's travel destinations and carefully evaluate the risks. But seniors are among the more durable and knowledgeable of travelers, and some travel on strenuous excursions well into their nineties. If you haven't done so, try it, you'll like it.

18. Welcome frustration

Robert Browning asserted, "A man's reach should exceed his grasp or what's a heaven for."[144] No sooner do we start to try to do something mental as well as physical, we are likely to come upon something at which we fail. Yet, the frustration of failure or near failure continues to inspire us to do better and try again.

I spend most of my retirement years writing on history, politics, and social issues. I do not have an agent and dealing with publishers is frustrating. But I continue to read, research, and succeed in getting my manuscripts published. In the process, I have learned to endure, even welcome, rejection letters, which form the bulk of my daily mail. But, Browning again:

"Then welcome each rebuff
That turns earth's smoothness rough,
Each sting that bids nor sit nor stand, but go!
Be our joys three parts pain!
Strive, and hold cheap the strain;
Learn, nor account the pang; dare, never grudge
the throe!"[145]

19. Do some good

You can do some good in the world by giving what you can to charity. You need to choose your charities carefully, because many of them do not spend the money as you designate. Some have such high overheads that only a small portion of what you donate are given to those who need help. More important than giving money is making a commitment to be helpful to a major project in which you are interested. Organizations that try to help the homeless, or those that seek to promote greater literacy, or those that help administratively in organizations devoted to Alzheimer's disease, Parkinson's disease, and AIDS, may welcome the help you can give.

20. Relearn to drive

Nearly all readers of this book have learned to drive. Young adults learned to drive by their parents, high school teachers, or by instructors in a driving training class. Reaching the age by which a child is eligible for a driving permit is something of a rite (or right) of passage for most teenagers. However, a new rite of passage is desirable for those over fifty-five. Reflexes may have grown stale, eyesight may have diminished, and new steering and braking devices on newer cars all suggest a refresher course could save your life and that of a pedestrian as well.

The American Association for Retired Persons identifies the following danger signals for elderly drivers:

*Feeling nervous or fearful when driving
*Having difficulty in staying in the appropriate lane
*Having too many close calls
*Getting into too many minor accidents causing some damage to your car
*Having difficulty in making judgments about the safety of passing another car or rejoining the mainstream of traffic
*Having other drivers honk at you frequently
*Becoming aware that many friends and relatives are reluctant to ride with you if you are at the wheel
*Getting lost frequently
*Losing peripheral vision so that you become unaware of pedestrians and vehicles at either side of you
*Having trouble seeing, obeying, and knowing the meaning of traffic signs and signals
*Responding too slowly to unexpected road conditions
*Having difficulty concentrating on driving
*Finding it hard to park in small spaces, or to be aware of cars behind you, especially when you change lanes or back up
*Getting warnings or traffic tickets
*Taking medications that may make you sleepy or otherwise impair your ability to react appropriately to driving situations

If you are experiencing a substantial number of these indicators, it is time to hang up the keys.

I gave up driving many years after I moved back to New York. For most seniors, giving up driving and giving up a car and a license seem a defeat, the march of time. The acknowledgment that one is older parallels the elation of a teenager who sees a new driver's permit as a deserved triumph. For me, however, it was a relatively easy decision. In Manhattan, a car is expensive to insure, expensive to park in a garage, and expensive to operate. Moreover, while costly, there is plenty of mass transportation and no end to luxurious travel by taxicab or limousine. I take relatively few trips out of the city, and so do not enjoy the countryside and the changing seasons as much as I used to. However, I have an up-to-date driver's license, which I use mainly as a form of photo identification. I could get behind the wheel any time if I wanted to. However, I will not do so before I take a refresher course from a reputable driving school.

21. Make a bet on the future

It doesn't have to be a big bet, but a bet on the future is an expression of confidence that a lot of life and laughter remain. If you have the opportunity, purchase a three-year subscription to a magazine rather than a one-year subscription. Assume that you will be around to enjoy each year, and it is likely that you will. If you can afford to do so, buy a new car. Shopping is fun. On a grander scale, remodel your home or apartment. Many, and perhaps most, seniors spend a lot of time at home. And while your apartment may be immaculate and neat as a pin, time and grime erode the original colors, and it may be time for a new paint job. Often the erosion goes unnoticed by those living with it, but it is there nevertheless. A new kitchen or a remodeled bathroom are expensive projects, but serve to offer a new perspective on life. It is up lifting, I think, to hold your ground with contractors until you get what you want at a price you want. Debate the color scheme with a decorator, if you have one, and with your children or grandchildren. But cling to your plans. It's still your home!

22. Know your limits

Every older person in the course of a full life will be a caregiver for a child, a spouse, a sibling. Such care should be administered to those you love and to those who love you, with skill, compassion, and devotion. But you also need to know your limits. Sometimes you cannot lift the person needing the care. At other times, you cannot assist with the more intimate chores of washing and toileting that patient. Under such circumstances, you need to find homecare professionals or place the patient in a residential healthcare facility.

When my father died, my mother fell apart. She required care for the fourteen years she lived beyond my father's death. The chores of dealing with the homecare workers, and otherwise administering to our mother's need, fell almost entirely upon my brother. The ailing person becomes first a source of compassion, but then is seen as an enemy. One cannot help but begin to wonder where are the golden years, the years of carefree travel, of weekends in the country, of dinners in fine restaurants, of time with one's own spouse and children? Know your limits and get help.

23. A senior moment of forgetting

This does not mean that we are about to come down with dementia or Alzheimer's disease. Memory loss is inevitable. Every reader of this book probably has a family member whose memory is so good that he or she would like to see him have at least a partial "forgetter!"

24. Invite friends for dinner

Asking a few friends for dinner enables the host to demonstrate cooking skills and to share those culinary skills with guests. Friends invited for dinner are an opportunity to exchange experiences and to establish an informal network of self-help. Good friends, fine wine, and a delicious meal go a long way in boosting mental health.

25. Read a few demanding books a year

These may be fiction or nonfiction books, but do not expect pulp fiction to exercise the brain. About six books a year, including biographies of women and men whose lives were inspiring to a generation of people, are among the best books to read. Good, well-told history and those books that provide detailed analysis of current events likewise fall into this category. Good journals of opinion may be challenging, as are good daily newspapers.

26. Build a repertoire of jokes

Few things are better icebreakers than a timely, well-told joke. You should memorize such funny stories so they appear spontaneous. Practice telling the joke so you can tell it with panache. Assure the funny story is timely to the group. Make sure the joke has bite, but also it must not offend the sensibilities of the listeners. You need to know your audience. How will they respond to off-color words or stories? Can they laugh at themselves? Can you laugh at yourself? Here's one that may be innocent enough to use:

Angel Gabriel was interviewing those who perceived themselves to be eligible to enter heaven. Angel Gabriel asked the first man, "Why do you think you deserve to be admitted into heaven?" The candidate for heaven responded by saying, "I was a man of God for forty years, and during those years, I preached the Gospel relentlessly and saved countless souls." Angel Gabriel replied, "That's very good, but step aside and we'll see." He went to the next candidate and again inquired, "Why do you think you should be admitted into heaven?" The candidate replied, "I have been a minister of the church for fifty years, and during that time, I made many converts and I think, therefore, that I am a good candidate for heaven." "We'll see," replied the angel. The third candidate Angel Gabriel interviewed was a bus driver. "Why should a bus driver be admitted to heaven?" he asked. The bus driver responded, "When your ministers were preaching, their congregants were sleeping. When I was driving, my passengers were praying." And with that, the bus driver entered heaven.

27. You can protect yourself against mistreatment

Among the crimes against elderly people, include those from paid or unpaid caregivers who steal their money, wine, liquor, jewelry, even cherished pots, and pans.

Another is from a friend or relative who hurts you. The latter is a particularly heinous offense often because the victim refuses to bring charges against a young relative lest it endanger his or her career. When a grandson or granddaughter steals, there is often a feeling that whatever was stolen will eventually go to him or her anyway, and so is not really a crime. Actually, it is and the police should be notified. Credit or ATM debit cards and other valuables should be hidden even from the most trusted of caregivers so that temptation is not thrust upon them, and because they are often under pressure from lawless relatives who urge the caregiver to steal from a wealthier invalid. "They won't notice it anyway." "Even if they do, they are so dependent on you, they will overlook it." Since the invalid is frail and probably unable to walk rapidly or at all, or may not be able to see or hear effectively, "You surely will not be caught." If you suspect you are being abused, call 9-1-1, your local police, or tell a trusted child or sibling about it. Above all, do not keep it to yourself.

28. A drink a day keeps the doctor away

Two drinks a day may keep the doctor away, but beware of drinking anymore. Recently, medical research has found that an alcoholic drink or two daily may have some beneficial effects. At least those who drink moderately have better longevity records than that of teetotalers. However, some people become easily addicted to alcohol, so exercise care and don't drink too much. If you can't handle moderation, then don't drink at all. If you take a drink, let someone else drive.

29. Write an essay summarizing your life to date

This is not only an intellectually stimulating activity, but by putting your thoughts and experiences in writing, you can see what you've accomplished, how much you have achieved, and how rich your life has been. Most people go through life taking one day at a time, which, I suppose, is okay. Yet, sometimes when one does that, each day goes by as "one damn thing after another." By putting your thoughts and experiences in the form of an extended essay, you can gain perspective of your life. You may include relations with children, relatives, in-laws, parents, and friends. You may describe how you felt if there was a falling out in a relationship, or how you felt when one died. If you can afford to do so, self-publish your essay. Have it nicely bound. Perhaps even distribute your piece to those about whom you have written, or share it with those you think might be interested and encourage others to write memoirs of their own.

30. Go birding from time to time

Put on some good walking shoes and a hat, slacks and a shirt with pockets, and go forth with binoculars in hand and see how many birds you can identify. You can keep track of the number of sightings for each species, and make notes where you first saw your feathered friends. To help with the identification of birds, Roger Tory Peterson's *Field Guides* is one of the best. Colorful pictures help identify common birds around your house, in the fields, and in the forests. As you become more sophisticated, you can begin to identify birds by the sounds they chirp. There are videotapes of birds and CDs on which you can learn their cries. One need not live in rural areas to enjoy birding. New York's Central Park is a prime area for bird watching. Whether it is in a city, a park, or a forest, take a friend and enjoy birding together.

31. Feed birds in your backyard

This is an aspect of birding, except that by feeding the birds in your yard, you encourage the birds to come to you. You may need some guidance in knowing what types of food attracts which birds. Keep bird feeders in such a place like your breakfast room so you and the birds can enjoy breakfast together. Beware, however, of the squirrels that seem to get into the most complicated feeder, and before long, it becomes a game of how to keep the ever-hungry squirrels at arm's length. An advertisement for a bird feeder with a plastic cylinder boasts it is a feeder that squirrels cannot rob. Alas, we discovered that particular brand could be broken into. Remember, also, that squirrels bite and some may be rabid, so take care with appropriate shots if necessary.

When I lived on Cape Cod in Massachusetts, birding developed into a long-term interest for my wife and me. There were three parts to our birding. We had several feeders, each with its own type of bird food purchased at the local feed store. I would fill the feeders about every other day and have daily jousts with the squirrels ably abetted in their crime by the chipmunks. Bird feeders need constant cleaning, and while this was a chore I would have rather avoided, the pleasure the birds gave me while watching them eat made the effort worthwhile.

A second aspect to bird watching on Cape Cod is to go to the shore early in the morning and/or at dusk to identify shore birds from various kinds of gulls, terns, and sandpipers. When the sun is down or not yet up, and the shore is still hard, you can see seagulls feeding off shellfish washed ashore. Just as one can get lost in the forest in pursuit of an infrequently sighted bird, so on a deserted shore, one can get lost as well, and so it is better to go to the shore in pairs.

Finally, several times a month, my wife and I would go to the Audubon Bird Sanctuary and there, deep in the forest with numerous signs giving directions as to destinations and distance, one can see and hear rare as well as common species of birds. On these birding excursions, we also learned to take our family. The grandchildren were fascinated at the birds they saw and the sounds they heard. Children tire easily, so it is best to plan for short excursions.

32. When you rent a car, go for it!

It will give you a lift, even if you pay extra (which you will), if you rent a car with some pizzazz. Perhaps a convertible or a snappy foreign number for taking those turns on the Cornish in Monte Carlo or in the rugged hills of the Italian Alps.

33. Light up
No, not a cigarette, but when you are walking along a remote road or highway or even on a city street, be sure drivers can see you. Wear something that reflects at night such as reflective gloves, buttons, or an armband.

34. Be canny about giving your money away
Elderly people are often prime targets for those seeking to raise money for allegedly good causes. While it may be better to give than to receive, whether you plan to give five dollars, fifty dollars, or more, do some research and be sure that the bulk of the money goes to the cause for which it was intended and not to line the pockets of administrators managing the program. Much of the money generously given to help the victims of hurricanes and earthquakes never reached the needy.

35. Pay off your credit card balances
With interest rates likely to go up, it is best to carry a zero balance on your credit cards. Do not charge more than you think you can pay off when the payment becomes due. Moreover, since most credit card companies notoriously charge penalties for balances that are even an hour or so late, give yourself plenty of time to send in your check or pay online.

36. Safeguard your important papers
It will not do to keep your important papers in a shoebox. What you need is a fire resistant and water resistant lockbox or minisafe. It is true that a thief may carry off some small safes and open them at leisure, but what is important is that at the very minimum, fire, water, or rodents do not destroy your important documents.

37. Music soothes the soul
"Listen to a soothing piece of music—gentle piano sonata, a mother's lullaby—and your pulse and breathing will generally slow down substantially." (*Consumer Reports* monthly newsletter, *On Health*. "Calming music to heal your mind and body." October 2006. P. 10.) According to this report, calming music reduces pressure and stress, encourages better sleep, lessens short-term and chronic pain, and eases Alzheimer's disease. Calming music can also "reduce aggression, restlessness, irritability,

wandering …. In some studies, music temporarily improved their memory and promoted positive social interactions."

38. Don't keep mistreatment to yourself

Mistreatment of the elderly is a growing concern. Most mistreatment comes from caregivers, whether they are paid employees or family members. Many elderly are embarrassed to admit that they are or have been the subject of mistreatment inasmuch as they are often dependent on the caregiver. Moreover, the elderly being fear placed in a nursing home if they complain about mistreatment. Mistreatment may be physical, verbal, or financial. But whatever it is that you fear or feel tell someone you trust. If you deem it necessary to go to the police, of course, do so.

39. Get enough sleep

The elderly often complain that they are unable to sleep, yet good health requires that the elderly sleep like a baby. The stress of life makes sleeping like a baby more difficult than it seems. Among the suggestions for falling asleep and getting the eight hours of snooze you need, the following Dos and Don'ts may apply.

If you have difficulties falling asleep, get out of bed and do something else. Avoid watching television because many programs offer more stimulation than may be conducive to sleep. Listening to calm music, as noted in Number 38, may be more beneficial than watching television. Avoid going to bed immediately after a big meal, and avoid going to sleep after drinking wine or spirits. Eat like a king early in the day and eat like a pauper in the evening. If you nap during the daytime, try doing so during the early part of the afternoon, rather than during the later hours of the day. Winston Churchill had the capacity of snoozing soundly for fifteen-minute intervals during the day and so gained the reserve energy to deal with the emergencies of governing the United Kingdom during wartime.

40. Safety proof your home

Take an inventory of your home to pinpoint potential hazards and make an effort to correct them before an accident occurs. Among the major sources of danger is fire from frayed and aging wires in lamps and electrical appliances, and water danger from defective plumbing. Be sure you know how to turn off the gas or water in an emergency and use care when lighting a fireplace or when burning candles. The fireplace should have a protective screen to keep the sparks contained, and lit candles should be away from walls and curtains. Make certain your smoke detectors work. Replace the smoke detector battery each time you change your clock. Remind yourself to do this when the time changes from standard to daylight savings time and then reverse the procedure. Have a sturdy stepladder readily available, especially in a fire or a hurricane emergency. Be sure your fire extinguishers work and you know how to use them. In the event of fire, keep doors closed to prevent the fire from spreading. Keep handy emergency telephone numbers for police, fire, and physicians. Floors should be non-skidding, and throw rugs (few in number), should be secure to the floors.

41. Ride a bicycle

For some elderly, riding a bicycle may be a safer means of transportation than driving a car. A good way to shop is to ride a three-wheeler to the store and bring your purchases home in the carryall provided. For the intrepid, *Consumer Reports* monthly newsletter *On Health* (Volume 18, No. 11, November 2006) makes the following additional safety recommendations:

*Protect your eyes from glare, bugs, and debris by wearing polycarbonate sunglasses.
*Position your handlebars so that your upper body and hands are comfortable.
*It is a good idea to wear gloves which absorb vibrations and can protect your hands if you fall.
*Attach a rear-view mirror to sunglasses, helmet, or handlebars.
*Adjust the seat so that it is level and that your leg can extend almost fully when the pedals are vertical.
*Always wear a helmet which should sit level on your head, not tilted back. The strap should fit snugly so that the helmet tightens when you open your mouth. Replace the helmet after an accident.

42. Don't stop exercising because you feel you are catching a cold

Use common sense. If you feel feverish or nauseated it is best to treat the common cold with bed rest, Tylenol, or some such medication. However, sensing the approach of a common cold, simply taper off a bit and see what happens.

43. Lick your wound

Licking a wound is almost a conditioned reflex to a sudden cut with a knife or other sharp object. While washing with soap and water may be a better solution, saliva contains bacteria-killing chemicals, which can help, not hurt, the cut.

44. Treat yourself to a manicure and/or a pedicure

This bit of pampering will raise your spirits, and keep your fingernails and toenails healthy. This is not a "women only" recommendation. Men, as well as women, go to salons that specialize in pedicures and manicures. The cost for such a visit is generally a lot less expensive than going to a spa. Be sure, however, that the person who treats you is appropriately licensed and the instruments are sterilized.

I was eighty-three-years-old when I mustered the courage to enter a beauty shop to ask for a manicure. A friend my age told me he had been getting a manicure for some time, and I thought I should too. I remember circling the shop two or three times to be sure there were few customers inside. I then took the plunge. My nails were clipped, cut, filed, and my cuticles pushed back. I declined the polish. I felt properly cared for after the manicure. For about ten dollars, plus a two-dollar tip, I felt the money well spent. Today, I go to the same beauty shop, greet the manicurists—some of whom I know—welcome other customers, and settle down for a relaxing fifteen minutes of pampering.

45. Coffee is a useless antidote to a liquor-induced hangover

The notion that black coffee can restore one's physical and mental disabilities brought about by excessive use of liquor is a widely assumed practice and is essentially a hoax. If you're drunk, don't drive. Lie down and rest if you can, and stop drinking. Let the alcohol take its course to wear off.

46. To fight a cold, chicken soup

Among the remedies that do some good and are still popular today are those that Grandma relied on during the good old days. Foremost is homemade chicken soup, which may relieve nasal secretions. To relieve a sore throat, tea with honey and saltwater gargles may be helpful.

47. Medical care abroad may be better than you think, and cheaper too

In order to save money or to get needed treatment for which insurance companies may be unwilling to pay, many Americans are treated in hospitals located overseas. According to the AARP Bulletin,[146] about half a million Americans are going overseas for medical care. The quality of such care is better than you might think. The costs are certainly cheaper, even if you include the airfare to distant locations. However, it is best to be cautious and know whether the medical treatment abroad is right for you.

1. Be sure you can communicate with foreign physicians in English or in a language in which you are both fluent.
2. The hospital, which you may find through a search on the Internet, should be accredited, and its accreditation should be carefully confirmed.
3. Be sure your domestic physician knows what you are considering. It is best to include her or him in your planning.
4. It is essential that you bring along your medical records as well as any medications you are taking. Continue to take the medication until advised otherwise.
5. A health travel agency will add to the costs, but agents will smooth the way for you, especially if you are essentially a stay-at-home person and have not traveled a great deal.
6. . Do not travel alone. A knowledgeable person can help if instructions and expectations are misunderstood.
7. Talk to the physician who will be treating you by phone, e-mail, or letter. The phone, at some point, may be the next best thing to being in the doctor's office. You can gauge through voice, vocabulary, or accent, whether or not communication will be a barrier.
8. When receiving treatment abroad, as here in the United States, mistakes can happen and medical procedures may fail or botch. Recourse, legal or otherwise may be difficult.

48. Testing saves lives—or does it?

Once upon a time, one went to a general practitioner; a family doctor now called an internist, to have a condition examined and given advice about what to do about the situation. In medicine, this is now a lost art. There are so many tests available for heart disease, skin and lung disease, and cancer. When there is a diagnostic test available, doctors are quick to administer it. They do so because they want to put the most sophisticated tools to work on their patients' observed abnormalities. They do so also, because failure to use the most advanced diagnostic techniques may bring about a lawsuit, and they do so because they want their patients to know they are at the cutting edge of medical care. Tests are often expensive, worrisome, painful, and may go wrong, or be misinterpreted. Pathologists can make mistakes, especially when they are tired as their day ends. Sometimes their "eyes glaze over" looking at an X-ray or magnetic resonance imagery. The least smidgen of a suggestion that cancer may be involved makes one leap to the most drastic surgical oncological techniques available. All of this is understandable. Medicine is as much art as science. The art of medicine lies in the determination of when to rely on intuition, when to rely on tests, and when to turn to the most advanced testing techniques. The wise doctor and the understanding patient recognize that in some situations, a period of less glamorous, watchful waiting may be the preferred approach.

49. Retire? Who? Me?

Americans are planning for retirement earlier but face the situation later. The reasons are many, but among them are the opportunities to tiptoe into retirement gingerly—to test the water as one would do, gradually immersing oneself into the ocean on a beautiful day. Many companies recognize that older employees often have irreplaceable skills to offer, and so make it possible for them to postpone full retirement by working part-time while retaining all or at least some of their job related benefits. If one can be flexible, it may even be possible to work, say, in a Wal-Mart in Massachusetts during the summer and in a Wal-Mart in Florida during the winter. Often the transition is eased by not requiring a new application for work at each location. This is not to say that all American companies are equally prepared to accommodate retirees, but many are doing so, and part of your retirement planning should be negotiating a gradual move from full-time, to part-time work. Management may be more flexible than you think.

50. No news is good news, think again

Often a physician will expect a patient to call about the results of a test on, say, a mammogram, or a PSA for prostate enlargement. Don't assume that the doctor will call you or that you will receive a call when follow-up procedures are needed. Most doctors are too busy to call you. Don't take a chance call your doctor!

51. Don't stop to play with or to admire a strange dog

Even when dogs are on a leash, they can bite, bark, growl, and otherwise show their displeasure. The dog walker may be inattentive, not know the dog well, or be unable to control its behavior among strangers. Pepper spray will keep an annoying dog at bay.

52. Some rules for preventing Montezuma's revenge

To avoid diarrhea while traveling in foreign countries, drink bottled rather than tap water. This rule applies not only when traveling in developing countries, but in advanced countries as well. The water in the former may be contaminated, but in the latter, even though the water has been treated, its chemical composition may be such as to give diarrhea to the unwary. The native has already built up a resistance to tap water in his or her country, but not so the tourist. Bottled water and bottled carbonated drinks are preferred. Brush your teeth using bottled water and avoid ice cubes.

53. Take the stairs

Walking up and down the staircase gives you the opportunity to incorporate aerobic fitness into your busy schedule. Walking stairs helps to burn calories, reduce weight, and improve the performance of your heart.

54. Be wary of therapy

Grief is the price we pay for the love we have for someone who has died. When grief reaches abnormal levels, many prescribe professional therapy. In most cases, however, even grief ends with time. While the period may vary from six to eighteen months, time heals and professional guidance from clergy or psychologists may be superfluous. The five stages of grief including denial, anger, bargaining (with God?), and depression, leads finally to acceptance. In normal stages of grieving, one never forgets and closure is a myth. Nevertheless, with acceptance of loss, we may open ourselves to others without fearing that we betrayed the deceased.

55. The paradox of medical technology

Medical technology that has the potential of helping us live independently longer may also may actually make us more dependent. That is, advanced medical practices may enable us to live longer, but these practices do not always restore us to a previous condition of robust health. Instead, we depend on others to administer pills, drive us to physicians, and see that we get the prescribed physical therapy. Who will perform these services? Can we rely on aging or even ill children to do so? Can we rely on professionals to care for us? If so, will we have the financial resources to pay for the services we require? A culture of mutual care beginning within the family may help one generation serve the needs of another.

56. Walk, don't run, to establish a course of medical therapy
When confronted for the first time with the catastrophic news of Alzheimer's, cancer, or heart disease, it seems natural to try to arrive at a concept for treatment as early as possible. This tendency is quite natural, since one may feel, "If only I had the cancerous lesion analyzed sooner I could have nipped it in the bud." Yet, it is more important to get the right treatment even if it takes a bit longer. Obtaining a second opinion from another physician is a vital first step. Help your physician by being knowledgeable about your condition. Read a book or surf the Internet for such information. Inform others, family and friends, about the dire diagnosis so they can be drawn upon to help you during the difficult times. Take someone you trust with you when you visit the doctor to be sure everything discussed is clearly understood.

57. Take a nap
Winston Churchill, the World War II prime minister of England, required eight hours of slumber. However, he often could not get eight hours of uninterrupted sleep, so he learned to take a nap or two during his long, twenty-four hour days. He could close his eyes for twenty minutes, sleep soundly, and awake refreshed. Not everyone can do this. But short, frequent naps can be invigorating, keep one alert, and help her or him welcome the daily activities that may lie ahead. Too many naps, however, those that are lengthy and frequent in duration, may be a symptom of some depression. However, in an otherwise physically and mentally healthy person, naps are a welcome respite during the day and serve to recharge one's failing bodily batteries.

58. Be wary of gossip
In any circle of friends, conversation generally revolves around the activities of other friends, their spouses, siblings, children, colleagues, and co-workers. In its best sense, gossip is a form of social bonding, which enables women and men to live amicably in social groups. Gossip is good because it contributes to the evolution of language, and the skills required communicating with one another. Gossip is friendly conversation. It can be thought of as the glue that binds people together. Of course, when gossip becomes malicious, that is, when gossip revolves around real, imagined, or potential scandal that harms careers and relationships, it is best avoided. In judging the character of another person, gossip may become more

important than observable facts. As Chico Marx asserted in the movie, *Duck Soup*, "Well, who you going to believe, me or your own eyes?"[147]

59. Healthy mind in a healthy body

Vigorous exercise on a regular basis helps to aid memory gain. Here again, the infirmities of growing older may make vigorous exercise difficult for the senior citizen. Test yourself daily with a new arm stretch, add a few more minutes on a stationary bike, or grab a slightly heavier free weight. One need not achieve the healthy mind in a healthy body as perfect specimens of womanhood and manhood see but achieving a level comfortable for you should be the goal.

60. Become computer literate

To survive in the twenty-first century, you need to understand the language of the digital age. One must learn to use a cell phone, send an e-mail, and use a BlackBerry. Before running about to find a teacher to help, first try asking your nine-year-old grandchild to assist you.

61. Enjoy guilty pleasures

*Those that enjoy an alcoholic drink or two a day have a lower heart attack risk than teetotalers. While two will do, three or four will not do at all.
*Chocolate is good for you. Dark chocolate, which is chock-full of antioxidants, improves arterial blood flow. Milk chocolate isn't bad either.
*After a big dinner, do not go to sleep for at least four hours to avoid possible heartburn pain. Thus, staying up late may be good for you.
*Get enough sleep to stay fit.
*Enjoy video games with your grandchildren; this will improve your visual and attention skills.

62. Fight clutter

In any well-lived life, we tend to accumulate or hoard a substantial variety of artifacts that punctuate the passage of time. It may take you the rest of your life in retirement to do so, but get rid of unused, unwanted, and outdated artifacts. Time, moreover, changes lifestyles, and much of what we accumulate takes up more space than the items are worth. Among these are wedding gifts, housewarming gifts, and those gizmos we no longer need, nor can we remember when or why we bought them. It is, however, difficult to part with memorabilia, especially when each has a story to tell, an episode to recall. Yet, for a new lease on life, the accumulator must give way to the thrower-outer. You'll discover that you don't need that larger apartment or home after all. The children you thought would bless you for leaving such a rich bounty of memories will cheer you for absolving them of the responsibility.

63. Beware of priceless antiques and original art

The eighteenth-century commode or that original Picasso may be precious to you, but your plans to bequeath them to children or grandchildren are doomed to fail. (See above.) It is not that these items have lost their value—the value may have increased. The new generation has their own ideas about priceless antiques, and their preferences in art may be different from your own. If your children or your grandchildren are into art and antiques, you may wish to give them that precious piece now, rather than wait for them to acquire such a possession in your will. If neither your heirs nor you prefer to keep the treasures, you may wish to sell them through a private sale or through an auction house. One can also donate the art and antiques to a charity and receive a tax deduction.

64. Obsessive-compulsive disorder is a preventable disease

If you cannot give away and cannot bear to sell your precious possessions of a lifetime, you may already be a victim of obsessive-compulsive disorder. Although the disease may be catching, all is not lost if you resolve to get rid of some of the stuff each week.

65. Become a nag

Husbands should pester their wives to get an annual mammogram. Wives should remind their husbands to get a digital rectal examination to determine benign prostate enlargement, or more frightening, prostate cancer. Spouses should also pester each other to get a colonoscopy every five years or so.

66. Ventilate your home

In order to save fuel, many people prefer to keep windows and doors tightly shut. However, it is better to let fresh air inside to reduce pollutants from fireplaces, carpets, upholstery, and paint that accumulate in stagnant air.

67. Move the medicine

Be sure your pills are in a cool, dry place. Because bathrooms and kitchens may not be suitable for such storage, place the medicine in a convenient closet. Check with your pharmacist to make sure the medicines you do not take daily are still effective.

68. Keep the periodontist away

A full two minutes of brushing your teeth twice a day, and a full two minutes of flossing a day will help keep the periodontist at bay. Easily said but difficult to sustain. Try it, but use a timer to get in your full two minutes.

69. Daydream

Take five or ten minutes twice a day to do nothing but breathe and meditate. Sit comfortably and close your eyes. It won't hurt if you even catnap a bit. Deep breathing may relieve stress, lower blood pressure, and increase energy output.

70. Combine exercise and work

For busy people, try combining work and exercise. One of the best ways to do so is to avoid the elevator and/or escalator and take the stairs. Despite a busy day, I frequently walk up the four flights of stairs to my apartment. Obviously, I would not do so if I lived on the twentieth floor. However, if you have many stairs to climb, try walking up the first few floors and then take an elevator for the remainder. The Empire State Building in New York City has an annual race up the stairs to its 100-plus floors. If I had a dog, I would walk my pet for the fun and exercise. If your chore is to mow the lawn, use a hand mower not an electric one. When you walk to work or for pleasure, swing your arms.

71. Read a lot

To be an interesting person you must read. Read more books—fiction or nonfiction, biographies, and autobiographies. Read informative magazines, journals, and the daily newspaper. Only print media can give you an in-depth analysis of current events. Television, irrespective of its numerous contributions to the enrichment of our lives, is fleeting. Only the written word can provide in-depth insights. Read for pleasure. In conversations, draw on information you gleaned from your readings to become a more interesting person.

72. Use it or lose it. Take time for sex.

Research suggests that sex is required for physical or mental health. Unlike cod liver oil, which my parents made me drink, sex is also fun. Intrusion by children, the ambition to earn enough money to pay the mortgage, the need to prepare meals, do the dishes, and wash the clothes, seemingly leave no time for intimacy. These essentials of life's routines, however, must give way to safe and consensual sex. More sex may contribute to better health in men, and may reduce the incidence of breast cancer in women. However pleasurable, sex may become routine and both parties must avoid letting this happen. Try or invent new sexual positions. Sex toys are useful but tricky; the battery may die at a crucial moment. A sexy movie may contribute to the libido, thus adding to energy renewal and recharging life's flagging batteries. Loving couples must avoid monotony by approaching sex with a sense of humor.

73. No fooling, no falling

Those who age gracefully must avoid falling. Jane E. Brody, health writer for *The New York Times*, quoted the National Centers for Disease Control that for people over the age of sixty-five, falls are the leading cause of injury-related deaths.[148] Joints stiff with arthritis can make it difficult for some elderly to recover from tripping. These falls can result in fractures of the hip, hand, elbow, or wrist. Some medications the elderly take can also contribute to imbalance and falls. Because falling often takes place in the home, the house must be clear of obstacles. Use non-skid mats in the shower. Use strategically placed handlebars in the bathroom to help get in and out of the bathtub and in and off the toilet. On stairs, install and use handrails and non-skid rubber tread. Keep the home well lighted. Much of this is common sense and requires that you know the pitfalls of mind, body, and home. You need to know where the booby traps are, anticipate an accident, and do something about it.

74. Remarry

With an increase in longevity, most married couples will find that when they reach the point of "'til death do us part," they might find themselves single. It is no sin and no sign of disrespect for the deceased if the remaining spouse remarries or otherwise maintains a long-term monogamous relationship. Married people are healthier and financially better off than those who remain single. Remarried husbands may come out better than remarried women, because generally, husbands live five years longer than those who remain single. Remarried men eat better, take better care of themselves, and drink and smoke less; remarried women show fewer signs of emotional distress. However, as Supreme Court Jurist, Harry Blackmun reminds us, "A wedding is an event; a marriage is an achievement."

75. Happy feet for a happy body

Most people wear the wrong size shoe. Many women wear high heels that are too high. This may make podiatrists happy, but irritation flows from those whose feet hurt. A faulty shoe size may cause both bunions and back problems.

76. Helmets for bikers and seat belts for drivers and passengers

As biking grows in popularity as a means of exercise and as part of the "green revolution," even older bikers will take to the streets. Accidents between a car and a bike are an uneven match, and the biker can improve his or her chances of being hurt by wearing a helmet. Wearing seat belts in the driver's seat or as a passenger in a private automobile or taxi will add an important margin for safety.

77. Prepare for an emergency

For Mormons, preparing for an emergency is a matter of religious observance. For the rest of us, it is rapidly becoming a lifestyle. Robert Hutchins, a brilliant university president asserted, "Get ready for anything because anything is what will happen." How much more so now in an age of terror played out on a background of economic downturn is an emergency plan needed—war is asymmetrical and civilians are in as much danger as soldiers. Terrorists are not, however, the only disasters for which we may need emergency planning. Natural disasters such as forest fires, hurricanes, tornadoes, blackouts, earthquakes, and floods, require emergency planning.

First, identify a safe place at home and keep a modest supply or ready cash in small denominations in the event of a global bank holiday. Thus, there will be some ready cash for groceries, gasoline, and other immediate necessities.

Second, keep non-perishable foods and water stored safely for immediate use for a three-day period. Make sure that emergency food does not require refrigeration.

Third, keep or develop a box of medical supplies for bandages and antiseptics, and another box for medications regularly required. Scissors would be useful, as would a battery-operated radio with shortwave capabilities so that you can hear what those in authority may have to say. One or more flashlights may come in handy, as would an extra supply of fresh batteries. The old reliable duct tape just might be useful to improvise a shelter.

Fourth, the American Red Cross recommends that at least one person in the family be trained in emergency matters. They also suggest that a family determine a place in the home as well as outside the home where they may rendezvous if the family separates.

It is not possible to plan for every emergency, but it helps to be prepared. Preparing your home for preservation in an emergency may be useful; do not make your house so impregnable that it becomes a prison.

Immortality, as you have read in this book, is probably neither a reasonable nor a desirable goal. Since death comes to us all at some point, the remaining items in this appendix are devoted to suggestions for meeting the end of life as serene and pain free as possible. Never say die, but ...

78. Prepare a living will

In a living will, you may spell out how you want to die. Do you want all possible measures taken to secure for you as many hours, days, or months you may have left? Do you wish to spare those you love—and who are near and dear to you—the agonizing decisions regarding the termination of your life? In a living will, you decide in advance the guidelines you wish followed. Advance health care directives, as these are called, are designed to provide guidelines for the treatment you want at the end of your life. Do you want a ventilator? Do you want intravenous feeding through a tube?

79. Designate a health care proxy

A health care proxy is a person who knows your wishes very well, and who you authorize to make your end of life choices when you can no longer do so. A health care proxy should be a relatively mature person. Sometimes older children will do just fine. Nevertheless, the decisions made must be altogether in your interest. Children may want to get you out of the way to speed the inheritance from your estate. Thus, a health care proxy should be a person who has nothing to gain from your demise.

80. Identify a legal professional

A lawyer will help you design a living will that conforms to the requirements of the state. If you have complete confidence in the lawyer, you may designate him or her as your health care proxy. Be sure that the health care proxy carries out *your* wishes, not those wishes of children or of others who love you.

81. Create a durable power of attorney
A durable power of attorney, created by an attorney, identifies a person who will control the finances of those grown old. It may be worthwhile to find a family member who has some skill in accounting and/or law to assume this role. It may even be wise to identify a second person to step should there be any suspicion that the first person may be misusing the account set aside for Grandma or Grandpa.

82. Prepare to take care of your parents
Whoops! Ninety-year-old Grandpa slipped in the bathtub and broke his hip. Who takes care of Grandpa? An adult child of, say, sixty may live far away and may be unavailable. Who pays for the expensive care Grandpa requires for both the short and long terms? A quick and dirty answer is to employ a health care professional. Since life spans are lengthening, a new science of geriatrics or caring for the old and the old-old is growing. A health care professional is one who can evaluate the situation objectively and is a caring individual who will not get emotionally involved in difficult situations. However, professional care is expensive when provided in the home, in a nursing home, or other assisted living facility. If one can afford to do so, hire a professional when Grandpa is well so the two of them may become acquainted and familiar. Discuss the care together that Grandpa may need in the future. Make it a team effort. To the extent possible, keep the elderly involved in community projects. Talk politics, art, music, sex, books, and sports. See to it that they go to lectures, plays, musicals, movies, and concerts. Encourage them to cast their vote. Widen the circle of their age-related friends. Sympathize with them when their peers die.

83. An aging mind is a terrible thing to waste
Aging minds may not react with the alacrity of younger ones. Older women and men may forget where they put their eyeglasses or respond to an argument more slowly than younger minds. Older minds may also take longer to catch on to a joke. Yet, we cannot do without the insights that aging minds bring. Mental activity remains critical if the elderly are to

enjoy their golden years. In an experiment, mental decline slowed among a group of aging beagles that were given a nutritious diet, the company of other beagles, and a lot of doggie toys as compared with those beagles that were not so fortunate.[149] Smart youth will grow up to be wise people as wisdom gains on speed. If dogs can do it, why not the elderly?

84. Part-time work—you may like it

In industry, older workers are perceived as less productive, not open to new ideas, and cannot keep up the pace that modern industry requires. However, as mental power replaces physical power, a place may be available for the retired worker considered too old to work. Moreover, since the older population is growing more rapidly than the younger population, industry may fall back on a pool of older workers. Since the customer pool is likewise growing older, many customers may feel more comfortable talking to a mature employee rather than a youthful one. Thus, wisdom may be a more valuable asset than mere youth and vigor. During a period of economic downturn, the older worker, who may prefer part-time to full-time responsibilities, may grow in number and in demand. The growing use of flexible work hours encourages the search for the willing, able, and older worker. The illegality of age discrimination in the workplace acts in favor of the retired or laid-off older worker.

85. Chicken soup is good for you

An untreated common cold may take a week to cure. A common cold treated with over-the-counter medicines may take seven days … so the old joke goes. Nevertheless, something in the old beliefs relied on homegrown remedies when little else was available. Mom or Grandma's chicken soup remains among the best exemplars of such beliefs. Tea with honey is good for temporary relief from a sore throat, as may a gargle of common table salt in warm water. Also, simply drink plenty of water. For a troublesome back and for a minor burn, use ice. Lick a wound.

86. Avoid or moderate your use of the hot tub, sauna, or steam room

Pampering yourself guilelessly is one of the joys of the twilight years. However, comfortable they may be, the hot tub, sauna, or steam room may be dangerous. For one thing, getting in and out of them may be tricky, because a wet floor offers a slippery surface upon which you may fall. So, be careful. If you soak over-long in a hot tub or sweat in a steam room or sauna, you may get dizzy. No more than ten to fifteen minutes, if that long, should be the rule; more than that and you run the risk of fainting. These comforting experiences may aggravate hypertension (high blood pressure), a common ailment among the elderly.

87. Beware: your tub and shower may be booby-trapped

Your bathroom may be an important contributor to home accidents. Wet porcelain fixtures, the tub, the shower, and the toilet may become slippery when wet and safety precautions must be taken. Among the safety essentials are the following: a rubberized, skid-proof mat in the shower, and conveniently placed handrails to assist you as you get into the tub or shower and on and off the toilet. A bath mat in the tub usually has small suction cups that adhere to the tub and make it a safe, relatively non-skid surface upon which to stand on when taking a shower, but not always. Be careful.

On one occasion, I checked into one of Milan, Italy's better hotels. Dusty from the journey from Paris, I decided to take a quick shower before unpacking my bags. I looked at the bottom of the tub to be sure a rubberized mat was in place. It was. I turned on the shower and confidently stepped in. To my shock and awe, the bath mat hydroplaned along the bottom of the tub, and I skidded with it. I was sure I had broken a leg or a hip, or perhaps both. With a good deal of effort, using the strategically placed handrails, I extricated myself from the tub. To my utter surprise, I did not get hurt—no broken bones. I could not help feeling that I had been the beneficiary of one of God's alleged miracles.

88. Take your vitamins

Among Americans, vitamins have become as ubiquitous as apple pie. *The Nutrition Business Journal* estimates that the industry will garner more than $9.2 billion in sales.[150] "Recent studies," however, "have indicated that taking a multivitamin won't protect you from heart disease or cancer."[151] While the claims made for taking vitamins may be overstated, if you feel it is beneficial to take them, you should continue to do so. Hey, you never know!

89. Smile; a young woman or man has just offered you a bus or subway seat reserved for the elderly

The first time a young, twenty-something rises and offers you a seat on the bus or subway, you feel mortified. There you were, caught with your age showing. Don't despair; be grateful. You know you want to sit, so do so with a thank you and a smile. The next time someone offers you a seat, it will be easier to accept the person's kind gesture. After all, you have arrived at one of the modest privileges allowed the elderly.

90. Move around to get around; it is never too late

It doesn't take much. A study done in Israel demonstrated that even four hours a week of physical activity might add years to your life. These activities need not be in the gym or around the track. Doing ordinary errands of shopping for food, browsing in a bookstore, or bringing and picking up clothes from the dry cleaners may be helpful in adding good years to the life of the elderly. Even an eighty-five year old couch potato will derive benefits from physical activity. The Israeli researchers assert that physical activity may be central in staving off the "onset of a spiral of decline."[152]

91. Do you blog?

A young friend of mine boasted that his four-year-old child was a master of the Internet. When I told him that I would be more surprised if he himself had mastered the Internet, and that I would be surprised still, if I, an eighty-something had done so as well. When I need to look up something, I turn to a reference book. When my great-grandchildren seek to look up something they turn to Google. Perhaps more than the addictive interactive games the computer makes possible, elders have become eager and engaged bloggers. Increasingly, they seem to impose their will on the blogosphere and master the art of cyberspeak. "Blogging helps keep older minds sharp, offers a platform in which to express views, and opens networks all over the world."[153]

92. Say it isn't so!

Elders have a reputation for being grumpy. Who can blame them? They are not only near the end of life but are probably frail as well. Yet, according to recent research, the alleged grumpiness of great-grandma and great-grandpa is more myth than reality. Youth are determined to confront a challenge and win. Experience has taught those in their golden years that most conflict will spontaneously go away. Youth are ready to grapple with conflict; the elderly know that most conflict will not last. Little wonder, then, when a grandchild goes to a grandparent with a problem, he or she may get little more than a hug and a pat on the head. This is not grumpiness but is an attribute of a contented old age.

93. Have a very Merry Christmas and all that

For many, the twilight years are lonely ones, sometimes mercifully interrupted by holiday visits of children, siblings, relatives, and friends. This could be a good occasion for more than a Yule log and eggnog. It could be an opportunity for a sober self-evaluation of your physical and mental health and further care. Use this time to share your thoughts about the end of life. This means not only providing details as best you know them, but of your financial condition as well. Is there a will? Where is it? Who is the lawyer who knows your financial status? Who is your accountant to help file income and estate taxes? Where are the insurance policies? What kind of final disposition of your body do you prefer? These questions need addressed forthrightly and need not blight the angst of a visit nor the good time during the holidays.

94. Marry in haste

Married people are less lonely, live longer, and enjoy the companionship of each other. Among the old-old, formal marriage, that is vows written on a piece of paper called a marriage certificate, may not be necessary. A long-term, monogamous relationship established by a meeting of minds and hearts may be enough. Such a relationship is often stronger than the paper provided by the state and the clergy testifying to the fact of matrimony. No prenuptial accords are required, children are more likely to be accepting of the relationship of Mom or Dad, and feel secure that neither partner has a claim on that portion of the parents' estate, which the children expect to be theirs in due course. Living in such a relationship is a good deal more flexible than formal marriage, yet is, by no means, the rather sordid relationship implied by the term shacking-up.

An author's tale. I have been widowed two times. My first wife, to whom I was married for forty-three years, died of breast cancer. My second wife, to whom I was married for six years, died of brain cancer. Today, I have a long-term (twelve-year), loving, relationship with a beautiful woman. I have offered marriage from time to time but her answers were evasive. So, no marriage. The ties that bind us are stronger than any vows we might take. If we wanted to disengage, nothing would be simpler. Yet, separation from one another is unthinkable. We will be together until sickness or death tears us apart from each other. That's the way it is!

95. As death nears, be kind to yourself and your family; consider hospice care

The Grim Reaper awaits us all, but the wait can be eased with palliative treatment. While hospice care offers no cure to patients with a terminal illness, it does offer a nearly pain free transition from life to death and offers comfort to the ones who love them. Hospice care is generally offered at home and for those who wish, spiritual counseling is available. Professional services of nurses, social workers, and physicians work as a team with the family. As Jane E. Brody, health writer for *The New York Times* asserts, "With hospice, death assumes a more natural trajectory, unencumbered by frightening machines and sometimes grotesque interventions of modern medicine and often make dying more painful for patients and families, as well as costlier for society."[154]

A principal of a large and prominent high school in New York and a close friend of mine had been ailing for some time. He called me one day to say that he had pancreatic cancer. I visited him at the hospital twice. One day, I received the following telephone call. "Jerry," he said, "I am calling to say good-bye. I am dying, and I am allowing myself to be admitted to Calvary Hospital." (Calvary is a superb hospice). Other than holding back tears, I could think only of his courage. Five days later, one of his children called to say that his dad had died. My friend was seventy-eight.

96. Go slow

A proverb of unknown origins goes something like this: "Be not the first by whom the new is tried, nor the last to cast the old aside." While it may be understandable for the aged to grasp the cup that contains the elixir of youth, there is also some wisdom in not being the first to imbibe new medicines based on hastily performed medical trials. Some big pharmaceutical companies that seek to get their new products out into the market quickly take experimental short cuts without giving new medications a chance to be thoroughly tested make the temptation worse. In medicine, being slow to adopt new medicines and new treatments maybe just what the doctor ordered. "Slow medicine shares with hospice care the goal of comfort rather than cure …"[155] A paradox of modern medicine is that as life spans grow, so grows the determination among the elderly to speed medical research and so grows the race to be among the first to try new medicines. Yet, medicines not fully tested may be among the most life threatening. A second paradox, which suggests that going slow, may be a more useful approach to speedy adoption of new medicines, is that tests to determine how effective a new medication may be are often quite painful and hazardous. In truth, the elderly want less of heroics and death with honor and dignity.

97. Medical marvels are on the way

Proton beam therapy can vaporize malignancies with little or no danger to nearby tissue. There are voice synthesizers for those whose cancerous voice box has been removed. One can avail oneself of artificial ears as well, if need be. Prosthetic limbs in various combinations can be made to order. Bionic men and women may not be around the corner, but they will be along sometime soon. Wired directly into the brain, bionic limbs may make the sedentary almost nimble. Organ transplants may become obsolete as engineers develop new livers and kidneys; there will be no waiting for kindhearted donors. Ailing organs may be repaired with stem cells. In the longer term, robots may replace surgeons. Sometimes surgeons make mistakes, or wake up on the wrong side of the bed the day you are scheduled for surgery. The robot is always at the ready and never errs! These medical marvels are not yet fully tested, all require extended periods of physical therapy, and one can expect some degree of pain as well. It is better to stay well.

98. Age is no country for old people

Relatively few physicians know how to treat old people. While stress may be the leading cause of early aging and premature death, few people know how to avoid stress and worse yet; physicians may not recognize it, and if they do, may not know how to treat it. Obviously, at some time or another, everyone is stressed about a job situation, experiences anxiety about one's health (or the health of parents or siblings), but what is important is to learn to deal with stress to lessen its affect while lengthening your life. While genetic factors play a part in determining how long or how well you will live, John Rowe, former CEO of Aetna, a health insurance giant, asserts that mainly "you are responsible for your own old age."[156] First things first. Lead a healthy lifestyle through nourishing meals, ample exercise, sound sleep, avoid smoking, and drink alcohol in moderation. Meditating, listening to good music, and practicing yoga, are among other activities one may undertake to reduce stress. Surround yourself with good friends and warm and fuzzy family. Do the things you like and think happy!

Work may well be a stress-reducing activity. I am never happier than when I am writing, organizing, or researching a new book. Even the rejection letters, of which there are plenty, are often a source of stress reduction to me, because I feel that at least I have been heard. The frustration comes from those many publishers who do not bother to respond in any way to my queries.

99. Are you prepared to make "A Gift of Life?"

Human life has been described as a pearl beyond price, but is it? In the 1950s, the era of transplant surgery may be said to have begun in earnest. "It is now within our grasp to replace nearly every and any diseased, or broken down, or worn out organ with a healthy one."[157] However, nearly everything has a price. In entrepreneurial America, surprisingly, the buying and selling of human organs is illegal. In America, a patient must find a donor willing to make the gift as a freewill offering and cannot exact payment in cash or anything of equivalent value. If the donor needs money and chooses to sell his or her kidney, is the gift of life, tainted by commercialism? Does it make the donor any less generous? The decision to donate an organ, a kidney, perhaps, is where the donor meets the scalpel, thereby assuming the risks of surgery.

The major argument against allowing organ donation to become a commercial transaction is derived from the laudable goal of giving the rich and poor alike an even chance of receiving a matching organ. However, since the buying and selling of organs remains illegal in America, does this put American patients at a disadvantage in the global market for scarce organs? In addition, is a black market in organs a better outcome?

Dr. Sally Satel asserts, "Altruism is not producing enough organs …"[158] The United Network for Organ Sharing, which has established an elaborate priority of organ donating, estimates that there are 100,000 people waiting for a transplant, and of this number, 9,000 die each year. If altruism is insufficient to shorten the wait for organs and if some lives can be saved by doing so, are you less generous for having contributed an organ and received money for it?

100. "Have you heard of the wonderful one-hoss shay?"

As described by Dr. Oliver Wendell Holmes in his classic folk poem, "The Deacon's Masterpiece," the one-hoss (horse) shay was built "in such a logical way" that it lasted a hundred years to the day and never broke down, it just wore out." Unlike the deacon's masterpiece, people are not created in such a logical way and they die by degrees. Biological systems break down, wear out, or are damaged or destroyed in an accident. Because death, like birth, is often a sloppy painful, affair, if you could choose a serene, painless drug induced death, would you do so?

101. To be or not to be?

In modern medicine, palliative or terminal sedation is medically possible. In America, however, doctor-assisted suicide is available only in the State of Oregon. Legal or not, many families are resorting to physician-administered painkillers that in the end, terminates life itself. For a loved one, a comfortable death may be preferable to a few more days or weeks of painful life. As the intravenous painkiller drips almost silently into the patient's arm or chest, life slips away in serene slumber. Because physicians take the Hippocratic Oath in which they pledge themselves to do the patient no harm, they are reluctant to admit that the practice is widespread. Shakespeare was ahead of his time when in Hamlet's famous soliloquy he warns us, "To die, to sleep perchance to dream: Aye there's the rub. For in that sleep of death what dreams may come when we have shuffled off this mortal coil, must give us pause." As Denise Grady writes in a piece for *The New York Times*, it is not so easy to get your physician to engage in a serious discussion of end-of-life matters. "It is a conversation most people dread, doctors and patients alike."[159]

102. Suicide? Don't even think about it!

When you start thinking about it, it may already be too late. Loss of the will to live is a common problem among the aged infirm with suicidal men far in the lead over women who may also be inclined. Older men living alone, lacking companionship, bereft of close friends, and feeling a purposeless retirement, are likely among the victims. Watching television and heavy drinking are no antidotes strong enough to combat the lost will to live. The ready availability of guns may make pulling the trigger a heroic way to go. Suicide, however, may not be the immediate blast from a gun. Instead, it may be a slow death in which men, and sometimes women, punish themselves by drinking heavily, dressing shabbily, eating and sleeping poorly, failing to take physician prescribed medicine, and otherwise cutting themselves off from their remaining family and friends.

Fortunately, symptoms of suicide abound. When a senior citizen expresses a view that his or her life is a waste, or that there seems to be no purpose in living, such expressions must be taken seriously. Engage in conversations, reminisce about the past, gossip, or talk about family members. Ask questions like "Have you visited your new nephew?" or

"How about joining me for a basketball game?" Buy the tickets and go! If you feel more help is necessary, encourage the senior citizen to seek psychiatric help. A suicidal tendency may well be a cry for help!

103. Old age is a never-ending adventure

Don't try this, but think about it. As described in *The New York Times* on January 8, 2010, a ninety-year-old woman sprained her ankle while hiking in South Africa. She did not quit after the mishap, but tried, unsuccessfully, to go on. An eighty-nine-year-old man's sense of adventure was such that he took up wing walking. He strapped himself to the wings of a single engine biplane, donned sun goggles, wore layers of clothing for warmth, stood on the airplane's wings, and flew across the English Channel. "My family thinks I'm mad,"[160] said the adventure-driven senior citizen. This eighty-eight-year-old author of this book agrees with the man's family. However, a sense of adventure may be satiated by reading a book on a subject you had not thought of before or had the time for in your busy and productive life. A sense of adventure may be sustained by trying a new restaurant, visiting a new museum, or by learning a new skill. More expensively, a sense of adventure may be encouraged by travel to new locations or revisiting places to which you had journeyed before. Risking the untried, the untested, or the untrammeled; maintain a sense of adventure and see the world at least once, or perhaps more than once, through the youthful eyes of grandchildren. These are some powerful ways of achieving at least a sense of immortality. Robert Browning said, "Dare never grudge the throe."

104. If it comes to that

Sometimes, at the end of one's life, it becomes evident that neither the family nor any one person in it can care for a frail, elderly person and a nursing home seems in order. For most families, placing a loved one in an impersonal environment, which reeks of a hospital not of a home, is a painful decision to make. Moreover, having decided that a nursing home is the only answer for elder care, which home to choose becomes a question that has destroyed more than one family. While costs are a factor—nursing homes are not cheap—it is most important that the home under consideration be visited. Do the residents seem content? Does the food seem wholesome? Is the staff helpful and caring? Is the facility clean? Do odors of urine or of unchanged linens linger? These are the observations one must make, and while a checklist may be helpful, the judgments are often neither intuitive nor objective. To keep harmony in the family, encourage others in the family to visit and make their recommendations.

105. A word to the elderly—be a mensch

Mensch is a Yiddish word that, casually translated, means a person of admirable characteristics. He or she is a person who chooses to be morally responsible for his or her actions. If you can say you lived with integrity, harmed no one else, was faithful to obligations at home, at study, and at work, you are a mensch. If you can describe yourself as a person who leans toward kindness and generosity, you are viewed as a mensch. Some see the world as it is and ask, "Why?" A mensch sees the world as it might be and asks "Why not?"

Appendix B

Profiles of Centenarians and of Other Late Bloomers

Despite its often-formidable challenges, life is worth living. How do we know? We know because the search for immortality is an ancient, not a new phenomenon. The Neanderthals sought immortality and the potential of immortality at or near the core of nearly every religion. Along the way, however, the search for immortality morphed from religious myth, to science fiction, and then to science fact. Janus-like, immortality faces the world of science fiction, and as the potential of living more years of productive life grows, immortality becomes a serious scientific endeavor. Because living forever in our own skin remains only a metaphor for humankind's growing longevity, we consider a random group of those who have made it to one hundred. Such is the yardstick by which, for now, we can take the measure of our days. But why stop at a hundred?

Grandma Moses (1860–1961)
"American Idol"
Following the death of her husband, Anna Mary Robertson Moses, "Grandma Moses," tried her hand at making embroidered pictures, but painful arthritis made it impossible for her to continue this detailed work. So at the age of seventy-eight, she took up painting, and in what seemed the twilight of her life, developed a talent that blossomed to make her world famous.

In 1939, Louis J. Caldor, an engineer and art collector, saw her work in a display window in a drugstore in Hoosick Falls, New York. He liked what he saw, bought the four paintings on display, and then drove out to her farm and bought as many paintings as he could. At age seventy-nine, Grandma Moses was discovered. Her paintings were included in an exhibition of the works of Contemporary Unknown Painters sponsored by the Museum of Modern Art in New York. Grandma Moses was no longer an unknown artist.

George Burns (1896–1996)
"I Can't Die, I'm Booked"

The audience loved it when he flaunted his age with the lines, "Nice to be here? At my age, it's nice to be anywhere I get a standing ovation just for standing."

In 1993, at the age of ninety-seven, George Burns was still a sellout. He was one of those entertainers who had to be on stage if he was to live. He preferred success to failure, but show business was the only life worth living.

Success did not come easily. Between 1910 and 1920, Burns acted in anything and everything, except the one thing that would make him successful. He danced, sang, and even worked with animals, but at age twenty-five, success eluded him until he became a comedian. But even as a comedian, he needed the support of Gracie Allen, an Irish Catholic actress, who was likewise looking for her main chance at success. Just why the audience loved her, Burns could not readily explain. In his 1988 book, *Gracie: A Love Story*, he put it this way, "Some kind of magical transformation had taken place. [The audience] loved her, I could feel it. It was the most amazing thing, and it happened just like that."

By 1925, the Burns and Allen show became theater, and for twenty-six years, a radio staple. George and Gracie married in 1926 and the couple moved to Los Angeles. There he joined the legends of Hollywood, including his best friend Jack Benny, as well as Groucho and Harpo Marx, George Jessel, Al Jolson, Eddie Cantor, Milton Berle, Edward G. Robinson, among others.

To the very end, George Burns flouted death. "I'm not interested in dying," he declared. "It's been done." George Burns died peacefully in Los Angeles just a month after his centennial birthday.

Amos Alonzo Stagg (1862–1965)
"A Century of Honesty"

The Grand Old Man, as students and faculty at the University of Chicago affectionately called him, is best remembered for his unassailable integrity and undeviating honesty in coaching and administering intercollegiate athletics. On his one-hundredth birthday in 1962, *Sports Illustrated* wrote a tribute to this athletic coach, which carried the title, "Amos Stagg: A Century of Honesty." However, Alonzo Stagg did not originally plan to be an athletic coach.

Instead, this shoemaker's son of West Orange, New Jersey, wanted to be a minister. After graduation from Yale, he began divinity studies. When he discovered that he was an ineffective speaker and would be a poor preacher, he joined the Young Men's Christian Association's Training School in Springfield, Massachusetts, as a student, coach, and faculty member. While in Springfield, William Rainey Harper, president of the University of Chicago, lured Stagg to the new university by appointing him as an associate professor and director of physical culture and athletics.

In 1900, he was named a full professor, and in 1914, the trustees of the University of Chicago named its athletic field Stagg Field, thus making it the first athletic field named in honor of a football coach. He reluctantly retired at the age of seventy in 1932.

Stagg, however, was not finished with life or with coaching. For fourteen years after retiring, he served as coach of the College of the Pacific in Stockton, California. In 1943, while still at the college, he was named Man of the Year by the Football Writers Association. He was on every American Olympic committee from 1902 to 1932 and was a track coach for the 1924 Olympic team.

He had made his mark not only as a football coach but also in developing rules of fair play and sportsmanship among the members of the National Collegiate Athletic Association. In 1958, he was inducted into the National Football Hall of Fame. Football Coach, Knute Rockne, spoke for football coaches everywhere when he said, "All football comes from Stagg."

Amos Alonzo Stagg was one of the founding members of the National Athletic Association. He pursued its mandate to protect the integrity of intercollegiate athletics, but with mixed results, attesting that the qualities Coach Stagg brought to the game were surely missed.

In 2005, a *Wall Street Journal* editorial declared, "Big time college sports are a mess."[161]

Lonnie, as Stagg was affectionately called, we need you still.

George F. Kennan (1904–2004)
"My Voice Now Carried"

George F. Kennan is best known for the "Long Telegram" to Washington, DC from Moscow. In that telegram, and in a lengthy article in the July 1947 issue of *Foreign Affairs* entitled, "The Sources of Soviet Conduct," which he signed with an "X"; George Kennan asserted that while Soviet power was impervious to the logic of reason, it was highly sensitive to the logic of force. This being the case, he urged that the expansion of the Soviet Union be contained short of war. By his own admission he declared, "My reputation was made. My voice now carried."[162]

George Kennan, widely viewed as one of America's "Wise Men," was born in Milwaukee, Wisconsin, on February 16, 1904 to Florence James who died two years later. At the age of eight, his father placed him in the care of his stepmother to learn German, the country that next to America, George loved the most. In a long diplomatic career, he mastered Russian, French, Portuguese, Czech, and Norwegian. He worked his way through Princeton University, which in 1925 granted him his baccalaureate degree. He said of his decision to try to enter the foreign service rather than return to Milwaukee as the "first and last sensible decision I was ever deliberately to make about my career."[163]

While his career at the highest levels of the diplomatic service was noteworthy, he felt more at home in academe. From the congenial environment of the Institute for Advanced Study in Princeton, he mingled with Albert Einstein and Robert Oppenheimer among others. In or out of diplomacy, he wrote many well-reviewed books and journal articles including *American Diplomacy 1900–1950* and *Russia Leaves the War* for which he won the Pulitzer Prize for History in 1957, the Bancroft and Francis Parkman prizes, and a National Book Award.

In 1989, President H. W. Bush awarded Keenan the Medal of Freedom, America's highest civilian honor. In 1974 and 1975, when Keenan was in Washington as a Woodrow Wilson scholar, he helped establish the Kennan Institute for Advanced Russian Studies. On that occasion, his modesty led him to recognize that the George Kennan Institute was named not only for one George Kennan, but also for an ancestor, his grandfather's cousin who bore the same name.

Gladys Tantaquidgeon (1899–2005)
"Live With Those You Love"

The venerated medicine woman of the Mohegan Indian tribe was one of seven children and a tenth generation descendant of Uncas, the heroic chief of the Mohegan tribe. It was through her leadership that the Mohegan's were recognized in 1994 as an Indian tribe by virtue of their tribal heritage, which she did so much to keep alive. She wrote several books on Indian medicine and folklore. *A Study of Delaware Indian Medicine Practices and Folk Beliefs* was first published in 1942 and reprinted in 1972 and 1995.

She continued her studies in anthropology at the University of Pennsylvania and awarded honorary doctoral degrees from Yale and the University of Connecticut.

Gladys Tantaquidgeon was called upon by numerous Native Americans to help them restore their tribal traditions and ancient Indian practices. She became sensitive to the needs of women, Indian and others, who had fallen on hard times. In 1931, in collaboration with her brother Harold and her ailing father, the tribe's former chief, she founded the Tantaquidgeon Indian Museum in Uncasville, Connecticut. She continued to work at the museum until she was ninety-nine. She did so in the conviction that, "You can't hate somebody that you know a lot about."

What follows is Tantaquidgeon's formula to live to be 106. Here's all one has to do:

Live with those you love.
Respect your past.
Honor the earth.
Have faith.
And, eat eight or nine very small meals a day.[164]
Mundo wigo, the Creator is good.[165]

Brooke Astor (1902–2007)
"You'll Have Fun, Pookie"

She was born Roberta Brooke Russell in Portsmouth, New Hampshire, on March 30, 1902. Her father was the sixteenth Commandant of the United States Marine Corps. Lest too much schooling spoil her for the life of a socialite, her parents married her off, ready or not, at the age of seventeen to a wealthy New Jersey socialite. Their son was born in 1924. The young couple; however, were not yet ready for marriage and much less ready for parenthood. Brooke's philandering husband was unfaithful and she obtained a Reno divorce in 1930. Her second marriage to Charles "Buddy" Marshall was happier, but he died all too soon. Now experienced in the one-upmanship game as played by American socialites, Brooke accepted the marriage proposal of Vincent Astor. She added new luster to the fading Astor name and worked hard to make her new spouse happy. Vincent Astor died at the age of sixty-seven. With a healthy trust fund of some $60,000,000, Lady Astor set out to do some good while having fun at the same time. As her husband, Vincent Astor predicted before he died, "You'll have fun, Pookie." She surely did.

Ms. Astor, unofficially described as the First Lady of New York, "moved effortlessly from the sumptuous apartments on Fifth Avenue to the ragged barrios of East Harlem, deploying her inherited millions to care for the poor." She became the victim of her son and his wife who, it was alleged, may have denied her care and comfort by stealing her favorite works of art, denying her prescribed medication, and forcing her to sleep on a filthy cot. This was hardly the reward the grand woman had expected. After a forty year tenure, during which she was the omnipresent link between two American Gilded Ages from the mid-nineteenth century immediately before the Civil War and the waning years of the twentieth century.

So integral a part did she become in New York's social and philanthropic environment that during one interview during the 1980s she described herself as a "public monument." She eminently deserved this title, because her hands-on involvement always followed her money, and the beneficiaries were the great museums, hospitals, universities, and libraries of the city. In 1977, the New York Public Library became the sole beneficiary of her fortune and her guidance. Her largesse extended also to the little people who worked anonymously to make the big people look good. Thus, no tablet appears to honor her for an air conditioning system, an employee's lunchroom, new windows for a nursing home, fire escapes for the homeless shelter, or a boiler for a youth center.[166]

Leni Riefenstahl (1902–2003)
"Where is My Guilt?"

"I must confess that I was so impressed by you and by the enthusiasm of the spectators that I would like to meet you personally." This is what Helene Berta Amalie "Leni" Riefenstahl wrote in 1937 to her hero, Adolph Hitler. Having been a well-known dancer and accomplished actress, her wish to meet the Nazi Führer was granted. Before long, there were numerous subsequent meetings, but probably no love affair. On one occasion, he told her, "You must make my films," and so she did. Thus, did Leni Riefenstahl put her genius for movie making at the service of Adolph Hitler.

Her originality in cinematography became evident when she made movies of the rallies of the Nazi Party in Nuremberg (*Triumph des Willens*) in 1934 and of the Berlin Summer Olympics in August of 1936. With a camera crew of 170, she filmed the Olympics in ways not heretofore conceived. While the filming of the Berlin Olympics was to be a Nazi propaganda event, she showed the victory of Jesse Owens, an African American, who won four gold medals in track and field and contradicted Nazi racial theories of Aryan superiority and the emerging "master race" of Nazis.

Leni was a tiny woman who defied aging. At age seventy-one, she took up scuba diving claiming to be but fifty-one she could get her diving license. She continued diving until late into her nineties. As an actress in films, she took dangerous roles that were ordinarily left to stuntmen. With bare feet, she climbed rock faces and once, was almost swept away by a director-created avalanche. In due course, she from movie making to photography, and at the age of ninety-seven, she went to southern Sudan to photograph the Nubian people.

While Lei Riefenstahl became disenchanted both with Hitler and the Nazi Party, she never completely denounced them, and in the public mind, she was always tainted and taunted by her early conviction that Germany's future was solely in Hitler's hands.

In 1993, when she was over ninety, the German filmmaker, Ray Müller, made her life the subject of a three-hour documentary, *The Horrible Wonderful Life of Leni Riefenstahl*. When Müller questioned her about feeling any remorse or guilt for her Nazi past, she assertively replied, "Where is my guilt? I can regret that I made the party film in 1934. However, I cannot regret that I lived in that time. No anti-Semitic word has ever crossed my lips. I was never anti-Semitic. I did not join the party. So where is my guilt?"

She was able to provide further insight into her life in her own 669-page autobiography, *Leni Riefenstahl: A Memoir*.

In a *New York Times* book review, John Simon noted, "The memoir did not contain a single "un-spellbinding page." Was what she remembered true? The reviewer replied, "The book must in the main be true, it is far too weird for fiction."[167]

Max Schmeling (1905–2005)
"Righteous Gentile"

In 1938, this now aging writer, gathered with friends around a radio to listen to the broadcast of what surely was the greatest heavyweight-boxing match in the world, then and now. In one corner of the boxing ring at Yankee Stadium sat Joe Louis, heavyweight champion, and an African-American upon whose shoulders rested the pride of the nation. In the other corner of the ring sat Max Schmeling, the German on whom the hopes that the greatest fighter would once again demonstrate the superiority of the Nazi ideology with its fanciful racial theories of Nordic invincibility.

Two years earlier, in 1936, Schmeling had returned to Germany to be hailed by the Nazi propagandist, all for defeating Joe Louis in twelve, hard-fought rounds. Joe Louis, for his part, had defeated James J. Barrack and had become the heavyweight champion of the world once again. However, Joe Louis insisted, "Don't call me champ until I beat Max Schmeling." Now he had that chance and he did not disappoint.

Schmeling had come to America for his return match, but there was no supportive parade for him. Because he was widely viewed as a Nazi sympathizer, he had to make his way through the stands to the boxing ring. The crowd threw trash and spat at him. Americans were in no mood to be "good sports."

White America and Jewish America held their collective breath. An eerie rare silence prevailed on the affluent Upper East Side even as it did among the less prosperous immigrants of the Lower East Side. That night, nearly all Americans were boxing fans. Those lucky enough to get seats at Yankee Stadium, and those who gathered around their blaring radios, felt the ultimate outcome in the imminent fight between democracy and dictatorship would be symbolically determined by the outcome of the fisticuffs between these two men. The match between these able, honorable, pugilist lasted exactly 134 seconds. The "Brown Bomber" as Joe Louis was often called, pummeled his opponent in the opening onslaught and the referee ended the fight.

This time there would be no triumphal return to Germany for Schmeling. After two weeks in Polyclinic Hospital in New York where his two broken vertebrae were treated, he was taken by stretcher for the homeward journey. Joe Louis had his revenge. Americans had their victory. African Americans briefly basked in the reflected glory of their boxing hero. Jewish-Americans were reassured that however bleak their future, they would ultimately survive.

American anger at Max Schmeling was understandable, to say the least. However, little did Americans know that Schmeling was not a political man. He would resist joining the Nazi Party, and he would not get rid of Joe Jacobs, his Jewish manager, as the Nazi hierarchy pleaded for him to do. He lived in Germany during the war years, served briefly in the German army where he was wounded, and then retired from public life. It was evident that during those years, he had not been a hard-core Nazi, and after the war, he would prosper handsomely. James Farley, President Franklin D. Roosevelt's Postmaster General, and later a highly placed executive at the Coca-Cola Company, awarded Schmeling the first Coca-Cola franchise for Germany. This made Schmeling a wealthy man. During the years when Joe Louis was frequently in financial need, his friend, and honorable combatant came to his aid.

It was well after the war when Jews discovered that on November 9, 1938, Kristallnacht, when many German Jews were murdered, and their businesses destroyed or taken from them, that Max Schmeling placed himself in harm's way by giving refuge to two Jewish children and earned their eternal gratitude.

Miep Gies (1909–2010)
"I am nor a Hero"

"I am nor a hero," wrote the last survivor of those who hid Anne Frank and her diaries from the Nazis. Miep Gies was being altogether too modest. In 1989, the West German government awarded Mrs. Gies its highest civilian medal, and in 1996, Queen Beatrix of the Netherlands conferred a knighthood upon her. What had she done to deserve such distinguished recognition?

For twenty-five months, Mrs. Gies, her husband, and three others hid Anne Frank, her father, mother, older sister, and three other Dutch Jews from the Nazis who were then occupying Amsterdam. Her memoir, *Anne Frank Remembered*, was published in 1987. Mrs. Gies remembered that on August 4, 1944, the Gestapo raided the back offices of Anne Frank's father, where the victims had been hiding. The journals that Anne Frank wrote described her life in the modest space that offered little privacy and little to keep the victims busy. Anne Frank, however, had used her time to make notes in a diary of her experiences in hiding. The Gestapo left behind these journals, sometimes only sheaves of scrap paper. Mrs. Gies was quick to understand their significance as first-hand testimony of how victims of Hitler dealt with the evil incarnate he represented.

Anne Frank died at Bergen-Belsen concentration camp three months before she turned sixteen. Her mother died at Auschwitz. Miep Gies gave Anne Frank's father, who had been liberated from Auschwitz, his daughter's journals to what later became, *Anne Frank: Diary of a Young Girl.*

Mrs. Gies was content to lead a modest, unpublicized life until an American writer encouraged her to write her version of those far-off but never to be forgotten days.

In *Anne Frank Remembered*, Mrs. Gies generously paid tribute to the other heroes of the Netherlands who risked their lives to save victims of the Nazis. "I stand at the end of a long, long line of good Dutch people who did what I did and much more."[168]

Thus did this Austrian Catholic survive and gave the lie to Holocaust deniers who can never seem to get it right.

Madame Chiang Kai-shek (1897–2003)
"The only thing Oriental about me is my face"

Through the long life of Madame Chiang Kai-shek, born Soong Mei-ling, the history of China during the last one hundred years may be studied. Her father, Charles Soong, was an American trained Christian missionary who worked closely with Sun Yat-sen, sometimes called the George Washington of China for his role in the overthrow in 1911 of Qing Dynasty, China's last.

Madame Soong was the youngest of three prominent sisters all of whom were educated in the United States at a time when women were not expected to receive an education, much less an education abroad, and certainly not in a Western country. Her two sisters married influential men in politics and banking, and her brothers held high posts in the Nationalist Party.

Soong Mei-ling married Chiang Kai-shek on December 1, 1927 in Shanghai and worked at his side when Sun Yat-sen died and power fell into Chiang's grip. It was he, as head of the anti-Communist Nationalists of China, who later fought, but was defeated by Mao Tse-tung, who eventually became the leader of the Communist People's Republic of China.

Madame Chiang Kai-shek was a perfect alter ego for her husband. She spoke fluent English, and the Generalissimo spoke English haltingly. He disliked holding discussions with foreigners; she readily reached out to them and helped persuade them to their point of view. As a representative in her husband's behalf, she tried to portray the nationalists as the only avenue by which China could free itself of tyranny, war, and poverty. In 1943, she became the first Chinese national to address a joint meeting of the Congress of the United States. Her appearance before Congress was a great success, and she obtained for her country millions of dollars in American aid and large additional sums from private contributions. She received many honorary degrees. She wrote *This is Our China* (1940), and *The Sure Victory* (1955).

When the Chinese Communists won a decisive victory over the nationalists, Madame Chiang and her husband were forced into exile in Taiwan, but they did not give up. Instead, they sought to bring America to recognize and support the Taiwanese government. They delayed Washington's recognition of Communist China for thirty years. In 1975, Chiang Kai-shek died and Madame Chiang,in recognition of the futility of her cause and her waning influence among the Chinese, left Taiwan and spent the rest of her life in semi seclusion between New York City's Gracie Square where she had an elegant apartment and a mansion on Long Island.

Soong Mei-ling professed democracy, but did not know its meaning and could not live it. The Soong family was believed corrupt and to have siphoned off a large part of the aid America gave China. Dazzling, energetic, arrogant, she was loved by some and hated by some; it was not as a compliment when she was given the name, Dragon Lady.

Claude Lévi-Strauss (1908–2009)
"I am not the father of structuralism"

Born in a family of French Jews, Claude Lévi-Strauss became the world's most distinguished and perhaps, the world's most controversial anthropologist. In his view, the myths of so-called primitive societies were not primitive at all. There was, he asserted, a universal structure behind all human activity. This concept of structuralism, while growing in acceptability among anthropologists, Lévi-Strauss is not without his critics. Yet, a French colleague says of him, "He is one of the intellectual heroes of the twentieth-century."[169]

In *Mythologies*, his four-volume work about the structure of mythology among Indians in the Americas, he sought to demonstrate that little-known myths were important to a better understanding of Indian cultural practices. For example, that cannibals boiled their friends and roasted their enemies, help in understanding the rites of passage among many Indian groups in America. While seeming to prefer the noble savage against the trained mind, Claude Lévi-Strauss was critical of each. Lévi-Strauss exalted the French philosopher Jean Jacques Rousseau and became a hero of the counterculture during the decades of the sixties and seventies.

Claude Lévi-Strauss was a man for all seasons. He took degrees in law and philosophy at the University at Paris, taught in a local high school, and became a professor of sociology at the University of Sao Paulo in Brazil. It was in Brazil where he made a commitment to anthropology, and with his wife Diana, whom he later divorced, began making trips into Brazil's interior.

In 1939, he returned to France to pursue the formal study of anthropology. During World War II, he was drafted into the French army where he served with British troops. In *Tristes Tropiques*, he describes the disorderly retreat from the Maginot Line. The judges of the *Prix Goncourt*, France's most prestigious literary award, sought to bestow its blessings on *Tristes Tropiques* because it was not fiction; however, they could not do so.

In 1941, Lévi-Strauss became a visiting professor at the New School for Social Research in New York. Later, he asserted that his experience in New York's Public Library—where he spent his time mostly in the library's reading room as "the most fruitful period in my life."[170]

In New York, he became part of an artistic circle of surrealists in painting and literature and spent a good deal of time and a lot of money buying what he called exotic antiques. He held a number of other distinguished positions. In 1973, at the age of seventy-five, he was elected to the French Academy.

Despite these achievements, structuralism remains a controversial subject among the world's *intelligentsia*. Claude Lévi-Strauss was a man of action and a man of thought. He died a month short of his 101st birthday still looking for "universal patterns, links, and modes of organization, and thought to find them, he would go anywhere.

Dr. Michael DeBakey (1908–2008)
"America's Greatest Surgeon?"

In 2005, *The Journal of the American Medical Association* asserted that Dr. DeBakey might have been "America's greatest surgeon ever."[171] His fame as a surgeon was such that he was entrusted to operate on Mohammed Reza Shah Pahlavi, the overthrown Shah of Iran. He also operated on the Duke of Windsor and the former King Edward VIII of England. He performed surgery on entertainers such as Marlene Dietrich and Jerry Lewis, athletic heroes such as boxing champion, Joe Louis and legendary baseball manager, Leo Durocher. Dr. DeBakey was consulted on the coronary bypass surgery of President Boris Yeltsin of Russia that prolonged his life while he contributed to a thaw in the cold war between the US and the former USSR.

Dr. DeBakey was surefooted walking the corridors of political power in Washington, DC as in the faraway capitals of nations large and small. Colossus-like, he straddled the world of medicine and politics, and by the scruff of the neck, held them snarling at one another, but eventually "reasoning together." To the chagrin of The American Medical Association, he supported the Federal Medicare program. President Lyndon Johnson named him Chairman of the Commission on Heart Disease, Cancer, and Stroke. He was among the first to focus the attention of the nation of the cancer-causing likelihood of smoking.

Dr. DeBakey's awards are too numerous to mention, even briefly, but we attempt do so nevertheless.

*In 1963 at the age of 55, he was awarded the distinguished Lasker Prize for "inaugurating a new era in cardiovascular surgery."[172]

*In 1969 at the age of 61, President Johnson awarded him, the Presidential Medal of Freedom—the highest honor America can bestow on its citizens.

*In 1974 at the age of 66, he was elected to the Soviet Academy of Medical Sciences after operating on Mstislav Keldysh, the country's distinguished nuclear scientist.

*In 1987 at the age of 79, President Ronald Reagan awarded DeBakey with the National Medal of Science.

*In 1987, he was also awarded the Congressional Gold Medal, the highest civilian Congress can offer. President Bush was at the ceremony in which Dr. DeBakey was so honored.

*In 1994 at the age of 86, he was still serving on the jury of the Mary Lasker Foundation where he supervised the awarding of high honors for medical achievement. Well into his nineties, Dr. Michael DeBakey continued to travel and speak at international conferences.

Dr. DeBakey had a big ego and the numerous awards did not exactly make him lovable. He was kind to his patients, and his Louisiana drawl was reassuring when they were prepared for surgery. While he had an almost unerring sense, he acknowledged that he was tough on interns. "If you were on the operating table," he declared, "would you want a perfectionist or somebody who cared little for detail?"[173]

Gordon Parks (1912–2006)
"I'm a rare bird"

He bought his first camera, a second-hand one, in a pawnshop in Seattle in 1938 and quickly recognized that a camera, creatively used, could be an important instrument to fight poverty, racism, and correct human wrongs. The photographic lens, as wielded by Gordon Roger Buchanan Alexander Parks, as he was named at birth, became a formidable weapon he brilliantly and provocatively used to photograph the portraits of African Americans such as Malcolm X (1963), the exiled Eldridge and Kathleen Cleaver (1970), as well as boxing and civil rights militant, Mohammed Ali (1970).

Although Gordon Parks had a talent for portraying the corrosive impact of poverty and racial segregation, he was equally able to capture in pictures the glamorous lives of the wealthy. He was a brilliant photographer of fashion models, film stars, socialites, and celebrities, including Barbra Streisand, Aaron Copland, Alexander Caldwell, Gloria Vanderbilt, Roberto Rossellini, and Ingrid Bergman. *Life's* photo editor assigned him to Paris where he photographed the funeral of the Nazi collaborator, Marshal Petain.

In a portrait entitled, "American Gothic," he photographed government housekeeper Ella Watson, standing unsmiling, and holding herself stiff and posing with a mop in one hand and a broom in another before an American flag. It was Parks way to express his anger at America where he had been refused service in a clothing store, in a movie theater, and in a restaurant.

He loved tennis and his yellow Jaguar and toured the country as a semipro basketball player. He wrote music, several novels, the most famous of which was *The Learning Tree* and a memoir, *A Choice of Weapons.* Nor was his photographic brilliance devoted to still pictures lone. He made a number of movies including directing the blockbuster film *Shaft* in 1971.

In 1943, during World War II, he was a correspondent for the Office of War Information. For twenty years, he was the only African American photographer for *Life* magazine when that journal was at the height of its popularity. Although he was largely self-taught, he was, by the age of fifty, one of the most versatile image-makers of the postwar years.

When *Life* shut down as a weekly picture magazine in 1972, it appeared that life was just beginning for this sixty-year-old man. In 1981, he wrote his second novel *Shannon* about Irish immigrants. In 1989, he wrote the music and libretto for the ballet *Martin*, a tribute to Reverend Dr. Martin Luther King Jr.

In 1987, a retrospective of his work was held in the New York Public Library and another retrospective at the Corcoran Museum of Art in Washington, DC.

In 1988, he received the National Medal of Arts award from President Ronald Reagan. Although Parks never finished high school, he was awarded forty honorary degrees from colleges and universities in the United States and England.

In an interview for *The New York Times* in 1997, Parks observed, "I'm in a sense, a rare bird. I suppose a lot of it depended on my determination not to let discrimination stop me."[174]

Suggested Additional Reading

John W. Rowe and Robert L. Kahn. *Successful Aging.* New York: Pantheon Books, 1998.

Larry Tye. *Satchel: The Life and Times of an American Legend.* New York.

Howard Chudacoff. *How Old Are You? Age Consciousness in American Culture.* Princeton, NJ: Princeton University Press, 1989.

Matilda White Riley. "Aging, Social Change and the Power of Ideas." *Daedalus,* Vol. 107, No. 4, Fall 1978.

Eric Cohen and Leon R. Kass. "Cast Me Not Off in Old Age." *Commentary,* January 2006, Vol. 121, No. 1.

Simone de Beauvoir. *The Coming of Age.* New York: Norton & Company, 1972.

Robert Butler. *Why Survive? Being Old in America.* New York: Harper & Row, 1975.

W. Andrew Achenbaum. *Crossing Frontiers: Gerontology Emerges as a Science.* Cambridge: Cambridge University Press, 1995.

Georges Minois. *History of Old Age from Antiquity to the Renaissance.* Chicago: Chicago University Press, 1989.

David W. E. Smith. *Human Longevity.* New York: Oxford University Press, 1993.

David Hackett Fischer. *Growing Old in America.* New York: Oxford University Press, 1977.

Gordon Moss and Walter Moss. *Growing Old.* New York: Simon & Schuster (Pocket Books) 1975.

Peter Laslett. *A Fresh Map of Life: The Emergence of the Third Age.* Cambridge, Massachusetts: Harvard University Press, 1991.

Thomas R. Cole. *The Journey of Life: A Cultural History of Aging in America.* Cambridge: Cambridge University Press, 1992.

Michael Fumento. *BioEvolution: How Biotechnology is Changing Our World.* San Francisco: Encounter Books, 2003.

Marvin Cetron and Owen Davies. *Cheating Death: The Promise and the*

Future Impact of Trying to Live Forever. New York: St. Martin's Press, 1998.

Sebastian de Grazia. *Of Time, Work, and Leisure.* New York: The Twentieth-Century Fund, 1962.

Ray Kurzweil. *The Singularity is Near.* New York: Viking, 2005.

Ray Kurzweil and Terry Grossman. *Fantastic Voyage: Live Long Enough to Live Forever.* New York: Rodale, 2004.

Joel Garreau. *Radical Evolution,* New York: Doubleday, 2004.

David N. Friedman. *The Immoralists: Charles Lindbergh, Dr. Alexis Carrel and Their Daring Quest to Live Forever.* New York: Harper Collins, 2007.

Notes

1. Quoted in John W. Rowe and Robert L. Kahn. *Successful Aging*. New York: Pantheon Books, 1998, p. 36.
2. Larry Tye. *Satchel: The Life and Times of an American Legend*. New York: Random House, 200, p. viii.
3. Howard P. Chudacoff. *How Old Are You? Age Consciousness in American Culture*. Princeton, NJ: Princeton University Press, 1989, p. 10.
4. Matilda White Riley. "Aging, Social Change, and the Power of Ideas." *Daedalus*. Vol. 107, No. 4, Fall 1978, p. 43.
5. Chudacoff, *op. cit.*, p. 117.
6. Quoted in Ken Dychtwald. *Age Wave: The Challenges and Opportunities of an Aging America*. Los Angeles: Jeremy P. Tarcher, Inc., 1989, p. 26.
7. Samuel Johnson. *London* (1738 edition) *Imitation of the Third Satire of Juvenal*.
8. Quoted in Howard P. Chudacoff. *How Old Are You? Age Consciousness in American Culture*. Princeton, NJ: Princeton University Press, 1989, p. 190.
9. Quoted in Eric Cohen and Leon R. Kass. "Cast Me Not Off in Old Age." *Commentary*, January 2006, Vol. 121, No. 1, p. 32.
10. Simone de Beauvoir. *Coming of Age*. New York: W. W. Norton & Company, 1972, p. 2.
11. de Beauvoir, *op. cit.*, p. 4.
12. Robert N. Butler. *Why Survive? Being Old in America*. New York: Harper & Row, 1975, p. 14.
13. *The New York Times*, January 31, 2004.
14. Robin Toner and David E. Rosenbaum. "In Overhaul of Social Security, Age is the Elephant in the Room" *The New York Times*, June 12, 2005, p. 1.
15. Quoted in Ibid.

16. See Landon Y. Jones. "Swinging 60s?" *Smithsonian*, January 2006, Vol. 36, No. 10, pp. 102–106.
17. W. Andrew Achenbaum. *Crossing Frontiers: Gerontology Emerges as a Science*. Cambridge: Cambridge University Press, 1995, p. 1.
18. John W. Roe and Robert L. Kahn. *Successful Aging*. New York: Pantheon Books, 1998, p. 4.
19. Nicholas Eberstadt. "Old Age Tsunami." *The Wall Street Journal,* November 15, 2005.
20. Ibid.
21. Ibid.
22. Ibid.
23. Ibid.
24. Adapted from *Living Better, Living Longer.* Harvard Publications: Vital National Statistics Reports, Center for Disease Control. *Wall Street Journal,* June 20, 2005, p. R3.
25. Rowe and Kahn. *op. cit.,* p. xii.
26. Ronald C. White, Jr. *The Eloquent President*. New York: Random House Inc. 2005. p. 3.
27. Quoted in Georges Minois. *History of Old Age from Antiquity to the Renaissance.* Chicago: University of Chicago Press, 1989. pp. 14–15.
28. Minois, *op. cit.,* p. 15.
29. Minois, *op. cit.,* p. 21.
30. David W. E. Smith. *Human Longevity.* New York: Oxford University Press, 1993, p. 13.
31. Ibid.
32. David Hackett Fischer. *Growing Old in America*. New York: Oxford University Press, 1977, p. 6.
33. Gordon Moss and Walter Moss. *Growing Old*. New York: Simon & Schuster (Pocket Books), 1975, p. 18.
34. Moss and Moss, *op. cit.,* p. 19.
35. Donald O. Cowgill. "The Aging of Populations and Societies." *Annals of the American Academy of Political and Social Science,* Volume 415, September 1874, p. 10.
36. Gabrielle Birkner. "New Safe Haven for Seniors" *Jewish Week*, June 3, 2005, p. XX
37. 37 "Safe and Sound: Protect Yourself and Your Loved

Ones From Being Mistreated." TIAA/CREF, Summer 2005, p. 8.
38. Irving Rosow. *Socialization to Old Age.* Berkeley: University of California Press, 1974, p. 12.
39. Peter Laslett. *A Fresh Map of Life: The Emergence of the Third Age.* Cambridge, Massachusetts, Harvard University Press, 1991, p. 97.
40. Quoted in Laslett, *op. cit.,* p. 98.
41. Ibid.
42. Quoted in Thomas R. Cole. *The Journey of Life: A Cultural History of Aging in America.* Cambridge: Cambridge University Press, 1992, p. 207.
43. Alexis de Tocqueville. *Democracy in America.* Vol. II, New York: Vintage Books, 1955, p. 4.
44. Fischer, *op. cit.,* p. 67.
45. Ibid.
46. Fischer, *op. cit.,* p. 67–68.
47. Minois, *op. cit.,* p. 306.
48. Robert Browning. *Rabbi Ben Ezra,* 1864.
49. For these observations, the author is indebted to Thomas R. Cole. *The Journey of Life: A Cultural History of Aging in America.* Cambridge: Cambridge University Press, 1992, p. 68.
50. Ibid.
51. Cole, *op. cit.,* p. 69.
52. Cole, *op. cit.,* p. 162.
53. Ralph Waldo Emerson. "Old Age." *Atlantic Monthly,* Volume X. 1862, p. 139.
54. Gruman, *op. cit.,* p. 32.
55. Alexander Leaf. "Every Day Is a Gift When You Are Over 100." *National Geographic,* January 1973, No. 1, Vol. 143, p. 92.
56. Dan Buettner. "The Secrets of Long Life." *National Geographic,* November 2005, pp. 2–27.
57. See Dan Buettner. "Living Healthy to 100." *AARP Magazine,* May and June 2008. pp. 49 and passim.
58. Belle Boone Beard. *Centenarians: The New Generation.* New York: Greenwood Press, 1991, p. 3.

59. See Belle Boone Beard for more on characteristics of centenarians, *op. cit.,* pp. 3–14.
60. Michael Fumento. *BioEvolution,* San Francisco: Encounter Books, 2003, p. 133.
61. Peter Laslett. *A Fresh Map of Life: The Emergence of the Third Age.* Cambridge, Massachusetts: Harvard University Press, 1991, p. 10.
62. Quoted in Osborn Segerberg, Jr. *Living to be 100: 1200 Who Did and How They Did It.* New York: Charles Scribner's Sons, 1982, p. 1.
63. "Living to 100: What's the Secret?" *Harvard Medical School,* pp. 1–8 passim.
64. David W. E. Smith. *Human Longevity.* New York: Oxford University Press, 1993, p. 122.
65. *The Wall Street Journal,* February 25, 2005.
66. John W. Rowe and Robert L. Kahn. *Successful Aging.* New York: Pantheon Books, 1998, p. 6.
67. Quoted in Osborn Segerberg. *Living to be 100.* New York: Charles Scribner's Son, 1982, p. 22.
68. Carey, *op. cit.,* p. 241.
69. Ralph Waldo Emerson. "Old Age." *Atlantic Monthly,* Vol. 12m 1863, p. 135.
70. See S. Jay Olshansky and Bruce A. Carnes. *The Quest for Immortality: Science at the Frontiers of Aging.* New York: W. W. Norton & Company, 2001, pp. 213–215.
71. Olshansky and Carnes, *op. cit.,* p. 101.
72. Quoted in Gerald J. Gruman. *A History of Ideas About the Prolongation of Life.* New York: Springer Publishing Company, 2003, p. 36.
73. Quoted in *op. cit.,* p. XX
74. Quoted in Ibid.
75. Quoted in *op. cit.,* p. 121.
76. Gruman, *op. cit.,* p. 123.
77. Ibid.
78. Quoted in Thomas R. Cole. *The Journey of Life: A Cultural History of Aging in America.* Cambridge, England: Cambridge University Press, 1992, p. 187.
79. Quoted in Cole, *op. cit.,* p. 189.
80. Ibid.

81. Osborn Segerberg, Jr. *The Immortality Factor.* New York: E. P. Dutton & Company, Inc., 1974, p. 85.
82. Segerberg, *op. cit.,* p.86.
83. Alexis Carrel. *Man the Unknown.* New York: Harper Brothers Publishers, 1035, p. 109.
84. For these observations, I have drawn mainly on Roy L. Walford, MD *Maximum Life Span,* W. W. Norton & Company, 1983, pp. 19–22.
85. Jonathan Swift. *Gulliver's Travels.* Oxford: Oxford University Press, 1999, pp. 225–226
86. Ibid.
87. See Leon R. Kass, MD. *Life, Liberty, and the Defense of Dignity: The Challenge for Bioethics.* San Francisco: Encounter Books, 2002. Especially Chapter 9, "L'Chaim and its Limits." pp. 256–274.
88. Kass, *op. cit.,* p. 268.
89. Osborn Segerberg, Jr. *The Immortality Factor.* New York: E.P. Dutton & Co. Inc., 1974, p. 253.
90. Quoted in Segerberg, *op. cit.,* p. 257.
91. Brian Alexander. "Stay Pretty: Introducing the Ultra Human Makeover." *Wired* Magazine, January 2000, Vol. 8 No. 1, p. 185.
92. Claudia Dreifus. "Live Longer with Evolution? Evidence May Lie in Fruit Flies." *The New York Times,* December 6, 2005.
93. Alexander, *op. cit.,* p. 186.
94. Ibid.
95. Joel Garreau, *Radical Evolution: The Promise and Peril of Enhancing Our Minds, Our Bodies, and What it Means to be Human.* New York: Doubleday, 2005, p. 5.
96. Quoted in Michael Fumento. *BioEvolution: How Biotechnology is Changing our World.* San Francisco: Encounter Books, 2003, p. 126.
97. Quoted in Fumento, *op. cit.,* p. 127.
98. *The New York Times,* December 27, 2004.
99. Ray Kurzweil and Terry Grossman, MD. *Fantastic Voyage.* New York: Rodale, 2004, pp. 3–4.
100. Quoted in Kurzweil and Grossman, *op. cit.,* p. 24.
101. Segerberg, *op. cit.,* p. 245.

102. Ray Kurzweil. *The Singularity is Near: When Humans Transcend Biology.* New York: Viking, 2005, p. 7.
103. Quoted in Kurzweil *op. cit.,* p. 14
104.
105. Quoted in Gerald J. Gruman. A *History of Ideas About the Prolongation of Life.* New York: Springer Publishing Company, Inc., 2003, p. 129.
106. Peter Laslett. *A Fresh Map of Life: The Emergence of the Third Age.* Cambridge, Massachusetts: Harvard University Press, 1991, p. 12.
107. Tara Parker-Pope. "The Secrets of Successful Aging." *The Wall Street Journal,* June 20, 2005, p. R1.
108. Laslett, *op. cit.,* p. 13.
109. Laslett, *op. cit.,* p. 14.
110. Laslett, *op. cit.,* p. 64.
111. *The New York Times,* July 14, 2005.
112. Laslett, *op. cit.,* p. 22.
113. *The Wall Street Journal,* January 7–8, 2006.
114. Quoted in Mark Benecke. *The Dream of Eternal Life.* New York: Columbia University Press, 1998, p. viii.
115. Ibid.
116. Nicholas Wade. "Your Body is Younger Than You Think." *The New York Times,* August 2, 2005, pp. 1–4 passim.
117. Eric Cohen and Leon R. Kass. "Cast Me Not Off in Old Age." *Commentary,* January 2006, Vol. 12, No. 1, p. 33.
118. Ibid.
119. Ibid.
120. Ibid.
121. Quoted in Sebastian de Grazia. *Of Time, Work, and Leisure.* New York: The Twentieth Century Fund, 1962, p. 381.
122. Quoted in Osborn Segerberg, Jr. *The Immortality Factor.* New York: E.P. Dutton & Co. Inc., 1974, p. xviii.
123. Andrew L. Haas. "A Crash Course for the Elderly." *The New York Times,* July 17, 2006.
124. 123   "Elderly Drivers: Not Ready to Give Up the Keys." *CBS News,* January 31, 2002.
125. Ibid.
126. Alan Pifer and Lydia Bronte. "Introduction: Squaring the Pyramid." Alan Pifer and Lydia Bronte (eds). *Our Aging*

Society: Paradox and Promise. New York: W. W. Norton & Company, 1986, p. vii.
127. Jacob S. Siegel and Cynthia M. Taeuber. "Demographic Dimensions of an Aging Population." Alan Pifer and Lydia Bronte (eds). *Our Aging Society: Paradox and Promise.* New York: W. W. Norton & Company, 1966, pp. 79–110, passim.
128. Siegel and Taeuber, *op. cit.,* p. 107.
129. de Grazia, *op. cit.,* p. 13.
130. Peter Laslett. *A Fresh Map of Life: The Emergence of the Third Age.* Cambridge, Massachusetts: Harvard University Press, 1991, p. 196.
131. de Grazia, *op. cit.,* p. 437.
132. Harry R. Moody, "Education as a Lifelong Process." Alan Pfizer and Lydia Bronte. *Our Aging Society: Paradox and Promise.* New York: W. W. Norton & Company, 1986, p. 200.
133. Robert Butler. "The Longevity Revolution—Gearing Up for the Granny Boom" *UNESCO Courier,* January 199.
134. Quoted in Ker Than. "Toward Immortality: Questions about Quality of Life Take on a New Twist." *Live Science.* MSNBC.com. May 23, 2006.
135. Quoted in Ker Than. "Would Life Extension Make us Less Human?" *Live Science* MSNBC.com. May 24, 2006.
136. Quoted in Ker Than. "Toward Immortality," Part 1: "Social Implications of Life Extension Debated." *Live Science* MSNBC.com. May 22, 2006.
137. Philippe Ariès. *The Hour of Our Death.* New York: Alfred A. Knopf, 1981, p. 605.
138. Robert Jay Lifton and Eric Olson. *Living and Dying.* New York: Praeger Publishers, 1974, p. 72.
139. Nicholas Wade. "A Pill to Extend Life? Don't Dismiss the Notion too Quickly." *The New York Times,* September 22, 2000, p. A20.
140. Nicholas Wade. "Searching for Genes to Slow the Hands of Biological Time." *The New York Times,* September 26, 2000, p. F1.
141. Brian Alexander. "Don't Die, Stay Pretty." *Wired* Magazine, January 2000, Vol. 8, No;.1 p. 180

142. Brian Alexander. "Don't Die. Stay Pretty," *op. cit.,* p. 181.
143. Brian Alexander. "Don't Die; Stay Pretty." *Wired* Magazine, January 2000, Vol. 8, No. l, p. 183.
144. S. Jay Olshansky and Bruce A. Carnes. *The Quest for Immortality: Science at the Frontiers of Aging.* New York: W. W. Norton & Company, 2001, p. 217.
145. Poem from *Andrea del Sarto* (1855)
146. From *Rabbi Ben Ezra* (1864)
147. Anthony Mecir and Katherine Greider. "Traveling for Treatment." *AARP Bulletin,* September 2007, Vol. 48, No. 8
148. Quoted in John Tierney. "Pssst: Facts Prove No Match for Gossip." *The New York Times.* October 16, 2007, p. F4.
149. Jane E. Brody. "Taking Steps So Aging Does Not Mean Falling." *The New York Times,* December 26, 2006.
150. See Bernadine Healy, "The Power of the Aging Mind." *U.S. News and World Report,* Vol. 143, No. 17, November 12, 2007, p. 66.
151. T*he New York Times,* December 5, 2009, p. B5.
152. Ibid.
153. "To Live to a Biblical Old Age, Stay Physically Active." *Health and Nutrition News Letter,* Vol. 27, No.19, December 2009, p. 1.
154. Quoted in Lee Roberts. "Elder Bloggers Stake Their Claim." *The New York Times,* April 11, 2006, p. 6.
155. Jane E. Brody. "In Hospice, Care and Comfort as Life Wanes." *The New York Times,* December 1, 2009, p. D7.
156. Jane Gross. "For the Elderly, Being Heard About Life's End." *The New York Times,* May 5, 2008, p. A.1
157. Quoted in Tara Parker-Pope. "The Secrets of Successful Aging" *The Wall Street Journal,* June 20, 2005, p. R1.
158. Gerald Leinwand. *Transplants: Today's Medical Miracles.* New York: Franklin Watts, 1985, p. 37.
159. Sally Satel, MD. "A Gift of Life With Money Attached." *The New York Times,* December 22, 1909, p. D5.
160. Denise Grady. "Facing End-of-Life Talks, Doctors Choose to Wait." *The New York Times,* January 12, 2010, p. D1.
161. Quoted in Kirk Johnson. "Seeing Old-Age as a Never-Ending Adventure." *The New York Times,* January 8, 2010, p. 1.

162. Skip Rozin. "The Brutal Truth About College Sports. *The Wall Street Journal,* September 15, 2005.
163. Quoted in Tim Weiner and Barbara Crossette. "George F. Kennan dies at 101." *The New York Times* (Obituary) March 18, 2005.
164. Ibid.
165. Peter Applebome. "Words to Guide a Life of Over a Century." *The New York Times.* November 6, 2005. p. 46.
166. Applebome, *op. cit.,*
167. Ibid.
168. Quoted in Alan Riding. "Leni Riefenstahl, 101 Dies." *The New York Times,* September 18, 2005.
169. Quoted in T*he New York Times,* January 12, 2010.
170. Edward Rothstein. "Claude Lévi-Strauss, 100, Dies; Altered Western Views of the 'Primitive.'" *The New York Times,* November 5, 2009, p. A28.
171. Quoted in Ibid.
172. Quoted in Lawrence K. Altman. "Michael DeBakey, 99. Rebuilder of Hearts, Dies." *The New York Times,* July 13, 2008, p. 1.
173. Quoted in *loc. cit.*
174. Quoted in *loc. cit.*
175. Quoted in Andy Grundberg. "Gordon Park, Photojournalist Who Showed Dignity Amid Oppression." *The New York Times,* March 8, 2006, p.C16.

About the Author

Gerald Leinwand PhD is President Emeritus of Western Oregon University. He was Founding Dean of the School of Education at Bernard M. Baruch College of the City University of New York He is the author of more than forty books including the widely used high school textbook *The Pageant* of *World History* (Prentice-Hall). He has also published 22 books on *Problems of American Society* (Simon and Schuster). Other publishers include Franklin Watts, Basic Books, Rowman and Littlefield.

In retirement he has studied medical ethics. He took a course on the subject at New York University, lectured to second year medical students at Cornell-Wei!. He has also written on social issues generated by organ transplants and by sophisticated artificial organs.

Dr. Leinwand has traveled widely, and as served as consultant in international education for schools and universities. In Oregon, he wrote a weekly newspaper column bearing the title *Musings* in which he commented on American and global world politics and on his travel observations. He is 89-years-old and lives in New York City.